ADVANCE COMMENTARY ON *THE GREAT CONNECTING*

Jim Cashel has identified our position on the precipice of potentially one of the great turning points of history. Almost half the world's population has never known the ability to connect with the other half and with each other. The result will change lives, economies, and existing social structures. Cashel describes how connections have consequences, and he explores both the promise and peril that accompany connecting three billion people living in societies built around the absence of such connections. *The Great Connecting* is a must read for those who want to get ahead of this transformation curve.

—Tom Wheeler, former chairman of the Federal Communications Commission and author of *From Gutenberg to Google: The History of Our Future*

Communications technologies are reshaping the planet—but we haven't seen anything yet. Three billion new internet users changes everything . . . and this wonderful book captures that change in a very human way.

—Jeanne Bourgault, President and CEO, Internews

Jim Cashel explores a truly consequential moment in history through a compelling interweaving of narrative, travel notes, and policy recommendations. As both a technologist and a physician, Cashel leads us from the tech campuses of Silicon Valley to the refugee camps of Southeast Asia. I highly recommend you join him on this readable exploration of *The Great Connecting*.

—Nicco Mele, Director, Shorenstein Center on Media, Politics, and Public Policy, Harvard Kennedy School

THE GREAT CONNECTING

THE GREAT CONNECTING

THE EMERGENCE OF **GLOBAL BROADBAND** AND HOW THAT CHANGES EVERYTHING

JIM CASHEL

RADIUS BOOK GROUP
NEW YORK

Distributed by Radius Book Group
A division of Diversion Publishing Corp.
443 Park Avenue South, Suite 1004
New York, NY 10016
www.RadiusBookGroup.com

For more information, e-mail info@radiusbookgroup.com.

First edition: June 2019
Hardcover ISBN: 978-1-63576-645-5
Trade Paperback ISBN: 978-1-63576-646-2
eBook ISBN: 978-1-63576-654-7

Library of Congress Control Number: 2019938143

Manufactured in the United States of America

10 9 8 7 6 5 4 3 2 1

Cover design by Erin Kirk New
Interior design by Scribe Inc.

For Guyla Cashel,
who demonstrates for the rest of us
the magic of curiosity

CONTENTS

INTRODUCTION

In February 2018, a rocket launched by SpaceX, the company founded by Elon Musk, placed a Spanish radar satellite called Paz into orbit. While the launch itself was uneventful, there were two significant stowaways on board: a pair of small prototype satellites, Microsat-2a and Microsat-2b, to test new communications technologies. SpaceX plans to use information from these tests to build a network of communications satellites called Starlink to provide broadband services across the planet.

There are currently about eight hundred functioning communications satellites in orbit, providing services across the globe. SpaceX has Federal Communications Commission (FCC) permission to launch 11,943 Starlink satellites in the next few years. The constellation is scheduled to be completed by 2025.

Starlink is just one of several large projects involving a new generation of communications satellites that will provide broadband services across the planet over the next few years. Other nonsatellite technologies, such as high-altitude balloons and solar-powered drones, are being developed by Google, Facebook, Airbus, and other major players to extend broadband to parts of the planet currently unreached by the internet.

In the twenty-five years since the World Wide Web appeared with the launch of the Mosaic browser, about half the population of the planet has gained access to the internet. In the next three to five years, the other half will be gaining access. That second half of the planet's population, notably, is the "poor half," with most of the three billion or so future internet consumers currently living on less than several dollars per day.

The impact of rapid internet extension in developing countries will be profound, since broadband will enable many previously unavailable services, such as information access, distance education, online banking, health services, and government programs. It will also, of course, bring many challenges, including fraud, misinformation, and hate speech.

The extension of the internet across the full planet, which I call The Great Connecting, is a momentous event in world history. Have there been any other global events that significantly affected billions of people over just a few years? Even considering wars, epidemics, famines, technologies, and religion, it is hard to think of any. The Great Connecting is doing just that, however, at this very moment.

Despite the fact that the connecting of the planet is so significant, it is, ironically, very hard to witness. It is happening in millions of simultaneous small steps, all essentially hiding in plain sight. It is happening in a kiosk in Harare, where a student is buying her first smartphone. It is happening in Redmond, where an engineer is designing new broadband satellite antennas. It is happening in Cuzco, where a nonprofit is teaching farmers to use a new app. It is happening in New York, where global policy organizations are setting telecommunications standards. It is happening in Kigali, where international firms are laying fiber optic cable. All are contributing their little bit in this process of connecting the planet—and the combined effect is profound.

It is also happening fast. While traditional internet technologies involving cell towers and smartphones continue to expand in developing countries, the new technologies involving satellites, balloons, and drones represent a major and speedy leap forward.

I decided to write a book about The Great Connecting. It is an epic story that I wanted to investigate and better understand. So I took leave from my technology company in California to spend part of a year traveling to regions with no broadband, including some of the poorest, most remote areas that are likely last in line for connecting. I also sought to explore regions just getting broadband and to speak with those most affected. I met with engineers and scientists driving the innovation of communications at some of the largest and most sophisticated technology firms on the planet. I spoke with policymakers who are thinking about the rules and implications of expanding connectivity. I sought out the most knowledgeable and passionate folks I could find who are leading the expansion of broadband.

Two questions animated my explorations. First, what are the implications, both positive and negative, of The Great Connecting in developing countries and for the planet overall? Second, and most important, what are the major players involved in connecting the planet doing to prepare, to best accentuate the positive and mitigate the negative effects of expanded connectivity? I entered the project with a hunch that many groups are sprinting toward wiring the planet, but few are thinking hard about the opportunities and challenges once that happens not many years from now. The dog is chasing the car—but what happens when it catches it?

In this book, I navigate through the current state of broadband across the planet, including who has it, who doesn't, and present trends. I explore the technologies that in the near future will have a tremendous impact on reaching billions of new people. I review many of the remarkable possibilities that global broadband will offer. I dive into a number of challenges of expansion—including some heart-rending stories that illustrate very real perils. I also propose a number of steps that governments, organizations, and individuals should be taking to best prepare for The Great Connecting. The pages contain narrative, travelogues, background primers, and policy recommendations, all in a mosaic that reflects the complexity of the global story under way.

Through my explorations, I came to view The Great Connecting as a complex relationship taking place: the population of half the planet is about to become closely connected to the other half for the first time. It is the story of a global union. Like any complex relationship, many aspects are at play: exploring what's possible, investing in growth, overcoming challenges, and discovering the best path to partnership. Those are the stages I describe in the book. I've even organized chapters along the path of this emerging relationship.

And like any complex relationship, there are aspects that are wonderful, and there are aspects that are terrible. The relationship requires investment and effort and an optimistic sense of the future. It also requires a clear-eyed idea of what troubles might lie ahead.

As it so happens, the story of this relationship begins in a land very far away.

ATTRACTION

I'm standing under a majestic acacia tree in Mutambi, Malawi. It is the end of the rainy season here, so the rolling hills of the tea plantations are a vibrant green, giving way to the rocky escarpments and waterfalls of Mount Mulanje in the distance.

I'm visiting rural communities in Malawi with the international nonprofit GAIA (Global AIDS Interfaith Alliance), which runs HIV prevention programs in remote regions of one of the poorest countries in the world. I'm surrounded by about two hundred Malawians, mostly young women, who have walked for hours to attend the weekly mobile health clinic here run by GAIA. Some have come from Mozambique, crossing the river that marks the border a few kilometers from here. GAIA staff have spread out. Some are interviewing patients and testing for malaria in a small community building they've appropriated for the day. Other staff are consulting on HIV in the back of the Land Cruiser ambulance. Other staff are triaging patients and weighing babies outdoors under the acacia tree.

I had asked GAIA's president, Todd Schafer, if he would mind taking me to "the end of the road." We're now here in Mutambi. It was a bit of a trek to arrive. I flew from San Francisco to Washington, DC, where I boarded a plane to Addis Ababa in Ethiopia. There I caught a flight to Kamuzu International Airport in Lilongwe, the capital of Malawi. The flight continued to Chileka International

Airport in Blantyre, Malawi. Then I went by pickup truck about two hours to the east into Mulanje District. I passed through Lujeri, finally taking dirt roads for about half an hour past the Mutambi open market to arrive at today's health clinic.

I don't know if this is actually the "end of the road," but it is a pretty good proxy for it. The small buildings in the community are tidy structures made of adobe bricks, but they have no electricity, no running water, and no refrigerators to store food. The indoor cooking fires are built from sticks from the nearby mountains. The small rooms are mostly free of furniture and other possessions. Villagers pump their water from a shared well. Clothes dry on short clotheslines outside.

The people here mostly are subsistence farmers, growing cassava, bananas, peanuts, and other staples. Between houses, some people are drying kernels of corn on plastic mats.

Some of the men travel to nearby tea plantations, spending full days picking or trimming the carpet of new leaves growing along the tops of the bushes.

I see a few hundred homes in the area, although the number is undoubtedly larger if one were to explore by foot. It's almost certain nobody in the community has a smartphone or computer or has ever even heard of the internet.

And that's true for the next community over from here also. And for the next district. And for the entire region. It's true to the east in Mozambique and to the west in Zambia. It's true across most of southern Africa.

There are millions of Malawians who barely have the basics. From a wider perspective, there are hundreds of millions of Africans, and billions of people across the globe, in these precise circumstances. They have never heard of Wi-Fi and have almost no exposure to technology.

It's hard not to look at the women gathered under the acacia and not be struck by the profound dichotomy of our age. They have so little. Others on the planet have so much. It's hard not to want to help.

I'm also struck by another dichotomy. They don't have access to information, communications, banking, health services, or education opportunities—but the phone in my pocket does. It can access all of that, effortlessly, when in range of a signal. There is a device in their midst that I take for granted but that they would likely regard as magic.

Rural Malawi represents one "end of the road." My home, the San Francisco Bay Area, represents the other. The Bay Area cafés are crowded with hipsters talking about AI, AR, and AWS (artificial intelligence, augmented reality, and Amazon Web Services). These are the people who are building technologies and web apps that are taking over the globe, and—although it is hard to fathom—they are knocking on the door of Mutambi. That sounds like an exaggeration, but there are six web apps that everyone in developing countries routinely adopts when coming online: Facebook, WhatsApp, Instagram, Google, YouTube, and Twitter. All six are headquartered in my little corner of the planet.

It won't be too many years before higher access to broadband and lower prices for bandwidth bring new cheap or free hotspots across Malawi, including in Mutambi. At least in theory, in a few years, the women under the acacia tree will have access to all the sophisticated online tools and services that we at my end of the road rely on heavily.

So what will inexpensive, reliable broadband look like in Mutambi?

Currently, people here don't know what the weather will be next Wednesday. They don't know the price of maize in Blantyre a couple hours away. My teenage daughters could answer both of those questions quickly from our home in California—but the women here under the tree are locked out. They don't have access to information on health, on education, on agriculture.

They have no way to save or transfer money. They have limited access to credit. They can't take advantage of distance education or telehealth. They miss out on many government services.

They also miss out on convenience. They can't send a quick message to a relative, coordinate a business meeting, or easily share information with the community.

All of these deficits will disappear with broadband.

It is true that if connectivity were to become available tomorrow, relatively few of the women standing under the acacia tree would be able to afford it at the start. Probably none would. But the health clinics in the region probably could. And some of the schools. And some businesses. And certain tourist destinations. Malawian business professionals would have access, as would most of the foreigners in the area. The region would start to become embedded in an information environment that currently doesn't exist. That itself is transformative, even before the women around me have direct access themselves. They may not have cell phones, but their schools and health clinics and businesses would all begin to connect and benefit.

Step by step, the price of phones continues to drop. Voice and data charges, which can be purchased in tiny increments, become more affordable. Access improves. More and more people see the value in phones, in terms of convenience, cost savings, or income. More and more people start to link up—as has happened in every other poor community in the world that gains connectivity.

Mutambi gives the impression that it hasn't changed a whole lot in the last decades (possibly centuries). There are few clues in the village that this is the twenty-first century, not the eighteenth. But in a few years, many of the deficits endemic to this rural region—information, health, education, finance—will disappear with inexpensive broadband. Mutambi will have the opportunity to make significant progress in many areas and directly improve the lives of those I'm standing with today.

I'm visiting a small regional hospital about ten miles from the women under the tree (i.e., a four-hour walk for the locals—everybody here walks). The facility is clean, organized, and quietly impressive. The medical staff are cordial despite being overloaded. The wards are full but currently have just one patient per bed, which by rural Africa standards is a light load.

The facility doesn't seem to have any internet access. The nurses' stations have no computers. I haven't seen anyone consulting a smartphone. I can see on my device that there is no Wi-Fi. The facility operates as an isolated island, apparently without meaningful communications connections to other health facilities or systems. It is doing its best, all by itself, with very limited resources.

Earlier we stopped by a local school. Hundreds of children, all wearing smart gray-and-green uniforms, were sitting outside, taking lessons from instructors who had erected blackboards under the trees. There weren't sufficient classrooms or desks at the school, let alone computers or internet access.

Even the dignified tea plantation estate where I'm lucky to be staying—a handsome vestige of colonial times—has no internet connection (or even a web page). I don't think of myself as an internet addict, but it surprises me how many times I reach for my phone for some purpose only to be reminded that I'm in the half of the planet where that maneuver doesn't work.

Welcome to Malawi—where most of the country is still broadband-free.

I confess that I didn't know anything at all about Malawi until recently. OK, I knew it was a country in Central Africa and that it was really poor. That's not a very impressive stock of knowledge.

Malawi is a small, landlocked Central African country wedged between Tanzania, Mozambique, and Zambia. Its eastern border is dominated by Lake Malawi, the ninth largest lake in the world (also known as the "calendar lake," as it is 365 miles long and 52 miles wide). At the end of the nineteenth century, the British colonized the Bantu peoples living here, and Malawi finally gained independence in 1964.

Malawi is one of the world's poorest countries. The International Monetary Fund ranks Malawi 182 (of 187) in the world in terms of per-capita income. The average income in Malawi is around $1,000 per year, but because some rich people live in the bigger cities, the great majority of the population, almost all of which is rural, lives on only a fraction of that.

Complicating matters, Malawi (and Central Africa overall) has been devastated by the AIDS epidemic. Worldwide, about thirty-five million people have died of AIDS. Most of those are in Central Africa. In Malawi, almost 10 percent of adults carry the virus (about ten times the US prevalence).

Beyond the statistics, you can see the effects of the AIDS epidemic in everyday life. In a school I visited, the principal's office had a small chart on the wall of the number of "orphans" and the number of "double orphans" per grade in the school. Since classes can reach 150 students per teacher, any given classroom is bound to have many orphans. Health workers in Africa even track a statistic called "orphan-headed households."

Despite the challenges Malawi faces, it is an appealing country to explore. The scenery is beautiful. The Great Rift Valley runs north to south through the country, providing stunning vistas and supporting an abundance of wildlife. Waters flow from the south end of Lake Malawi and wind toward the mighty Zambezi River.

The people are also lovely—polite, helpful, cheerful, and welcoming. The professionals I meet come across as smart, modest, and hard-working. The villagers in the poorer communities are enthusiastic and hospitable, and they enjoy a good laugh. In one village, the women greeted us with song and dance. Their chief, sporting smart blue slacks and a lime-green blazer, couldn't have been more gracious, inviting me repeatedly to come and stay in his village whenever I could.

I also confess to loving the place names. Districts near Mulanje, for example, include Zomba, Neno, Mwanza, Machinga, and Mangochi.

Those of us from rich countries typically don't have the opportunity to meet people from the other end of the road. Despite the fact that literally billions of people live in very poor circumstances, it's not as if you can just show up to a poor village in Malawi and start walking around talking to people. Even just getting here is really hard. So when you do have the opportunity, as I do currently thanks to GAIA, it feels like a great privilege. It also feels like a paradox. Is it possible that technology companies and initiatives far from Malawi will truly influence this place in the near future?

I'm sitting on an oak-studded hillside above Lompoc, California, a quaint coastal community that bills itself as the "Flower Seed Capital of the World." The sun has recently risen behind me, and I see fingers of fog reaching into the valley below from the Pacific Ocean in the distance. The only sound is from a rooster on an outlying farm.

Suddenly rising from behind a ridgeline to my left is a brilliant orange column of flame, bright as a second sun. Atop the column is a SpaceX Falcon 9 rocket, launched from nearby Vandenberg Air Force Base, accelerating toward orbit. In about ten seconds, the roar from the nine engines reaches me, rumbling like sustained thunder.

This morning, SpaceX is launching ten communications satellites for Iridium, the communications firm that manages an aging fleet of sixty-six satellites. (The firm originally intended to build seventy-seven satellites, the number of electrons orbiting an Iridium nucleus, but ended up with only sixty-six, yet it decided not to rename the company "Dysprosium.")

Iridium is launching sixty-six new "state-of-the-art" low-earth-orbit satellites to provide voice, data, and broadband connectivity for the entire planet. Costing about $3 billion, the project is billed by the firm as "one of the biggest tech upgrades in history." The company has launched fifty satellites to date, all using SpaceX rockets, and there are twenty-five still to launch. (Iridium plans to put nine "spares" in orbit along with the sixty-six functioning satellites.) The entire constellation should be complete in the coming months.

It's all a remarkable show of technological prowess—rockets, satellites, global communications. And I'm thinking—it is amazing that this morning's spectacle may someday soon influence the women of Mutambi.

Traditionally, communications satellites have cost hundreds of millions of dollars to design, test, and deploy. Costs now, however, are plummeting thanks to more powerful technology, miniaturization, and cheaper launch capabilities. Lower costs per satellite mean many more can be deployed economically, which in turn allows orbits much closer to earth. Communications satellites closer to earth provide faster, higher-bandwidth services.

Much of the focus on new communications satellites is to support the demand of rich urban areas. My city of San Francisco will soon enjoy much better satellite communications. Those same satellites, however, will also be passing directly over Mutambi.

A major reason for cost savings is new, cheaper launch services. And global launch services are plummeting in price mostly due to one company: SpaceX.

In 2001, Elon Musk was thinking hard about Mars and the need for humans to become a multiplanet civilization. Few others that year were particularly interested in Mars; even NASA had no meaningful plans involving the planet.

Musk wanted to craft a publicity stunt to stir interest in Mars. His idea was to launch a greenhouse of plants to the red planet. Images of life growing in the distant environment would surely spark enthusiasm on earth.

When he enquired about launch expenses, however, he was dumbfounded to learn that a rocket would cost at least $60 million. He even negotiated with the Russians to use a decommissioned ICBM missile. When that didn't work out, Musk decided to found his own rocket company, which he named SpaceX.

Today, SpaceX is the most active launcher of rockets on the planet with about thirty scheduled launches per year, which is about a quarter of all launches globally. Its market share has grown quickly, in part because it has focused on lowering costs through reusability. SpaceX has mastered returning the first stage of its rockets to earth with precision landings, sometimes on land and sometimes on barges at sea—tiny targets in an ocean of blue. Its Atlantic Ocean barge is named *Of Course I Still Love You.*

SpaceX launches communications satellites for other companies now, but it has plans to start launching its own satellites—a lot of them.

There are currently around eight hundred functioning communications satellites in space. SpaceX has received permission from the FCC to launch nearly twelve thousand satellites total in the future. The firm has already launched trial satellites to initiate the program.

But SpaceX is notoriously secretive about its plans. It is required by law to submit applications to the FCC and other government agencies about certain initiatives, but outside of that, it has revealed relatively little except the name of this program, "Starlink." The usually talkative Musk says little about the program.

If things go according to plan, however, Starlink satellites will be providing broadband services to even the most remote communities on the planet. How will those services impact the lives of the poorest?

I'm in the outskirts of Tipitapa, Nicaragua, behind a ramshackle hut, admiring the work of Nicaraguan laborers constructing a well. The land here is dry, and there is no piped water, so residents depend on wells. The water table has been dropping in part because a major landowner nearby grows huge fields of rice requiring lavish irrigation. His wells are more powerful than other wells in the region, so extracting water for others is a major struggle. From where I stand, the region appears parched. It's hard to believe that if you dig deep enough, you may hit water.

Because the land is so dry, the wells—which in this poor region are dug completely by hand—need to be between 100 and 150 feet deep to function. I peer into the current excavation, which is about a meter wide, and mumble, "You've got to be kidding." It's like I'm looking down from the top of a fifteen-story building. Way, way down in the far depths, I can see the shimmer of some water and hear a chisel pounding against a distant stratum of rock. Somebody is hard at work down there.

Well digging isn't a safe occupation. Sometimes the projects cave in. Sometimes the earth releases toxic fumes. Sometimes there isn't enough oxygen. Occasionally the bottom suddenly falls out, sucking the worker down farther. (That is why he's always roped in.)

Because of the effort and the danger involved, wells are extraordinarily expensive, costing between $1,000 and $2,000 for the least complicated ones. That is a king's ransom in these parts.

One of the other workers around me, who occasionally tugs buckets of rock and slurry to the surface, tells me he heard recently about a team of young well diggers in the region who are using an entirely different approach to constructing wells, involving much narrower holes (just eight inches across) and using pipes, water pressure, and (if I've understood the Spanish correctly) the struts of old cars. I can't really visualize it, but apparently the method is faster, cheaper, and much safer than what I am witnessing here.

"How did they learn how to do it?" I ask.

"They saw it on YouTube," he replies.

Which technologies are currently reaching the poorest people on the planet? I come from Silicon Valley, where technology is viewed as the answer to every problem. But in Malawi, the only technology I see in the poorest regions is a bicycle (and occasionally a simple cell phone). As I currently bounce around the dusty rural roads of Nicaragua and speak with community members, I'm curious. Which technologies are appearing in their communities?

The answer depends on a community's economic status. As I drive through neighborhoods, it is relatively easy, even for a foreigner, to estimate the economic status of the residents by looking at their homes. By extension, one can estimate the degree that technology might impact their lives.

At the most impoverished level are shacks pieced together from scraps of corrugated metal, wood, and plastic that look barely habitable.

Families with more resources acquire custom-cut sheets of corrugated metal for their walls and roofs and essentially live in metal boxes.

Economic progress may bring an electric line connecting the house, which signals that the inhabitants can afford a dollar or two per month in electric bills. The power probably allows for a lightbulb or two and perhaps a fan.

Next will appear a makeshift antenna on the roof, indicating the presence of a TV, often a surprisingly early purchase for poor families.

In wealthier districts, the walls start to transition from corrugated metal to other materials—cinder block, stucco, wood. Perhaps a second room will be stuck on to the back. Residents who can afford such house upgrades almost certainly own a basic-feature phone, which in this region costs around five dollars. Call usage costs a few more dollars per month.

Next comes a satellite dish for cable access, which extends television viewing options from a few distorted channels to a hundred clear ones. Cable service starts at less than twenty dollars per month. It seems incongruous to see a new, shiny red satellite dish perched on a dilapidated metal roof.

Then other appliances will appear. In the houses I visit, I often see a fan and a small refrigerator. The ramshackle abodes will grow larger and sometimes even balance a second-story room that looks vulnerable to the next strong wind.

If folks own a satellite dish and a refrigerator, they probably also own a smartphone, although the cheapest phones and plans are expensive in this context—around fifty dollars for the phone and at least twenty dollars per month for calls and data. Even in relatively remote areas across Nicaragua, phone and data coverage appear pretty good (I frequently check by trying to pull up the *New York Times*), but data charges that are de minimis for me are out of reach for the folks in the houses I'm visiting.

The people I'm seeing in the region, and the levels of economic status they represent, are proxies for others in poverty across the planet. If a household can afford anything beyond subsistence, and if there is access to electricity, a lightbulb and a cell phone are very early purchases. It is a fairly big step to other appliances and smartphones.

A Quick Primer on Global Poverty

To understand the populations most likely to be impacted by The Great Connecting, a bit of background information on global poverty might be helpful.

"Extreme poverty" is currently defined as an individual living on less than $1.90 per day. By this benchmark, in 1990, 37 percent of the world's population was living in extreme poverty. Today, that figure has dropped dramatically, now approaching 10 percent. The United Nations (UN) identifies the first of its "Sustainable Development Goals" as the elimination of extreme poverty by 2030.

Much of the progress in reducing extreme poverty is thanks to efforts in five countries: China, India, Indonesia, Pakistan, and Vietnam. These countries alone moved an astonishing seven hundred million citizens out of extreme poverty between 1990 and 2010, partially through direct government programs but mostly due to overall economic growth. Unfortunately, in Africa during the same period, the number of people in extreme poverty rose from 290 million to 414 million people as populations soared and growth lingered.

The task of eliminating extreme poverty gets progressively harder the closer we get to zero poverty. Those still in extreme poverty are generally in rural or remote areas and lack electricity, sanitation, transportation, internet, or other fundamental services.

While the number of people in extreme poverty is dropping, it is important to remember that most of the planet is still extremely poor. About half of the planet lives on less than $2.50 per day.

If that is the story with global poverty, what is the situation with global wealth?

In 2017, Oxfam produced a report, "An Economy for the 99%," which included a staggering statistic: the richest eight men in the world have as much wealth as the bottom 50 percent of the planet (representing around 3.6 billion people). That's correct—eight people with a combined wealth of around $425 billion have as much as the bottom 3.6 billion people. (For those keeping track at home, the eight richest men in early 2017 were Bill Gates, Amancio Ortega, Warren Buffett, Carlos Slim Helú, Jeff Bezos, Mark Zuckerberg, Larry Ellison, and Michael Bloomberg.)

While all eight of the richest men could fairly be described as "self-made," that isn't true for the majority of billionaires (nine out of ten of whom are men), according to Oxfam's analysis. Over half of the world's billionaires inherited their wealth or work in industries prone to cronyism and corruption.

Eighty-two percent of the new wealth created in 2016 went to the top 1 percent of people on the planet.

A major impediment to economic development is a lack of access to electricity. More than a billion people have no access to electricity. Many more have highly unreliable service. Fortunately, new technologies are making major inroads in communities around the world to address the dearth of electricity.

The falling cost of renewable energy has allowed for the wider adoption of small-scale solar power. Small household systems—which typically include a solar panel, battery, lights, fan, and a cell charger—are getting both better and cheaper. Paying for systems is also getting easier. New "paygo" financing schemes allow users to pay for their systems each day rather than paying fully up front. In Kenya, for example, the company M-KOPA installs household systems and allows users to pay monthly. If a payment is missed, M-KOPA can turn off the system remotely until payments resume.

In India, about one-fifth of the population—over 250 million people—have no access to electricity. That should change soon. The government has committed to providing every home in the country that currently lacks electricity with a "microgrid" (a solar panel, battery, five LED lights, a fan, and a power plug).

The impact of these sorts of programs is hard to overstate. When a household has even modest, reliable access to electricity, kids can study at night, there is less indoor air pollution from kerosene, and communications, job opportunities, and public safety are all enhanced. It also heralds the coming of broadband.

I'm walking through a shantytown in Soweto, South Africa. The dwellings, each about half the size of a shipping container, are made of dirt and corrugated metal. The narrow paths between the homes are hard-packed and tidy. Most homes have entire families living inside of them and feature just a mattress, an area to cook, and not much else. I don't see any signs of plumbing or electricity anywhere.

Soweto, best known as the heart of the resistance to Apartheid, comprises several dozen townships southwest of Johannesburg ("SOuthWEst TOwnships"). The townships range from relatively affluent communities that include attractive single-family homes to shantytowns like the one I'm visiting. One relatively prosperous street in Soweto boasts the former homes of two Nobel Peace Prize winners—Archbishop Desmond Tutu and Nelson Mandela.

There are about three million residents of Soweto overall. My guide today tells me that only one resident is white—a woman married to a Sowetan who runs a bike touring company with her. An exaggeration? Perhaps. But in my exploration of the area, I'm thinking my guide may be right.

At the far end of the shantytown, near the road to Johannesburg, my guide points out the web of illegal wiring from several homes reaching up to a power pole nearby. One wire extends directly to the house of his neighbor, who offers a cell-phone-charging service for a small fee. While residents here don't have running water or electricity, they do have cell phones. Smartphones aren't popular because they are too expensive, but I see from mine that the signal is good, so the infrastructure is in place. It is just a matter of time before handsets and data plans become cheap enough to reach the poorest in Soweto.

How cheap do smartphones and data plans have to become before those at the bottom of the economic ladder can start to afford access? How much are people of limited means willing to pay? There is a running debate about this among economists, telecommunications companies, and phone manufacturers. But the answer typically is that they are willing to pay "more than you might expect."

I get a glimpse of that demand on a warm evening in a place far, far from Soweto.

EXPLORATION

It's late on a Monday night in Havana, and the narrow street to my hotel is unlit and so dark I can hardly trust my next step. In most other cities of the world, I'd be uncomfortable, but I'm not here. Crime in Havana is very low because of ubiquitous security cameras and serious consequences for petty theft.

As I approach my residence, I see several dozen locals sitting on the stonework adjacent to my hotel, their faces aglow. Each is staring intently into a cell phone. Havana has truly terrible Wi-Fi coverage, and the tourist hotels offer some of the better "Wi-Fi islands" in the city.

The web surfers' attention strikes me as particularly intense. Perhaps it is because access to the internet, provided by the state telecommunications monopoly ETECSA, costs about $1.50 per hour. That's a fortune here. A teacher in Cuba makes about $15 per month. A physician makes around $40. I know the Cubans aren't thinking in terms of $1.50 per hour. They are thinking in terms of spending a Cuban Peso every two minutes. They are focused on getting every second of value from their airtime.

I have no idea what is capturing everyone's attention—email to relatives in Miami? Airbnb account management? Cat videos?—but I'm reminded of my own willingness to pay for bandwidth when it is scarce. I remember the huge AOL bills in the 1990s for (by today's standard) slow and limited service. What exactly were we accessing then, anyway? My internet bill when I lived in Quito, Ecuador, years ago was about as much as my rent.

There is a running debate about what exactly the billions of people on the planet living on less than a few dollars per day will use the internet for—and whether they can afford to access broadband at all. We'll soon be finding out. I think these glowing faces on a warm evening in Havana offer an answer.

What Is Broadband?

The term *broadband* typically refers to an internet connection that is always on and has a high connection speed. If people have "broadband," they have persistent access to the internet. In this book, I also periodically use "bandwidth," "connectivity," and "Wi-Fi" to mean persistent access to the internet.

In developed countries, the initial internet connections in the early 1990s were not persistent. Typically, people linked to the internet through dial-up modems at slow speeds. Those connections allowed for email and a few other information services but were quite restricted.

Broadband arrived in the late 1990s and was typically offered either over phone lines (digital subscriber line, or DSL) or through cable service. People then started connecting through local Wi-Fi base stations, so their computers, and later other mobile devices, were always connected to the internet at high speed.

Simultaneously, mobile phone providers started adding data capabilities, beginning with snail-slow 2G (the "second generation" standard allowing some limited data services) but then progressing through 3G, 4G, LTE, and now in some regions 5G (with each generation greatly increasing data speeds and reliability). These mobile data connections became critical with the introduction of the iPhone in 2007, which offered feature-rich mobile access to the internet for the first time. Since the iPhone arrived, a slew of competitors has appeared. No matter which smartphone people have, they now take for granted that it is always connected.

Developing countries have skipped over most of this story. Initial access for new users is primarily wireless, accessed through cellular networks via smartphones.

Broadband (as opposed to intermittent access to the internet) is useful for consumers (think streaming video and web apps like Uber) but is vital for businesses. Cloud-based services, distributed databases, remote access to resources, mobile apps, and pretty much everything else a business does these days demands reliable, persistent connectivity. Broadband quickly evolved from a luxury to an absolute requirement for essentially all business and commerce in developed countries.

What percentage of the planet currently has affordable access to broadband? The answer depends in part on specific definitions. Different organizations use different data connection speeds in defining broadband. The FCC, for example, defines broadband as exceeding speeds of 25 mbs downstream and 3 mbs upstream. The International Telecommunication Union (ITU), UN, and others use lower benchmarks.

According to the UN's State of Broadband 2018 report, the best estimate was that 49.2 percent of the world's population would be online at the end of 2018 with reliable, affordable broadband access. Regions obviously vary greatly: Europe is 80 percent online, Africa only 22 percent.

The percentage of people with simple cell phone coverage that allows voice calls (but no data) is much higher than those with internet access. Simple cell phones (also referred to as "feature phones") are now cheap enough and accessible enough that most regions on the planet are now connected by cell phone. There are currently about seven billion cell phones on the planet, about the same number as humans on the planet (although penetration obviously varies greatly—from 240 phones per one hundred people in Hong Kong to fewer than 10 in many regions of Africa). According to a recent

Facebook study of seventy-five countries, 94 percent of the overall population had access to 2G networks (which are sufficient for voice and texting), while only 76 percent had access to 3G (data) networks or better—and many of those networks are still very expensive to use.

Simple cell phones are widely available. This is a remarkable achievement, representing a truly consequential global technology.

It also represents the leading edge of the internet: once people have simple phones, it is really only an issue of cost and affordability—and in some cases, expanded broadband coverage—to start moving to more capable smartphones. And the transition to smartphones, which about half the people on the planet are now going through, is the true game changer. It's convenient to be able to call and speak with somebody with a feature phone, but having full access to the information and services afforded by smartphones represents a major opportunity for those coming online.

The biggest current challenge confronting the expansion of global broadband is that most of the regions not yet covered are rural and poor. It is often prohibitively expensive to lay fiber optic cable (or any cable) to rural regions. Cell tower coverage is easier but requires a critical mass of paying customers to make the economics viable. Cell towers are generally placed one to two miles apart (at the least). They can be expensive to build, sometimes demanding new road construction and often requiring diesel generators for power. The fixed costs of cellular infrastructure limit the regions cellular networks can serve.

To complicate things further, the next generation of cellular technology is 5G, which is optimized for rich countries and cities. It allows a huge number of high-speed connections (in anticipation of the "Internet of Things"—where everything is hooked up to the internet), but it is very expensive to deploy. Gartner estimates that twenty million "things" will be connected by 2020, with millions more to follow. The bandwidth needs for self-driving cars alone will be enormous. My music system, my air conditioning, my toaster—all may require their own bandwidth soon enough, as enabled by 5G. That's all great for rich countries, but 5G isn't designed at all for poor, rural regions. It's too expensive.

So for now, only half the planet has smartphones. Most of the planet does have simple phones—and that in itself is a big first technological step toward The Great Connecting.

I'm in a learning center in Dario, Nicaragua, speaking with the staff about cell phones.

Dario is a simple town that is infrequently visited by outsiders. Its first guesthouse only recently opened. Dario's claim to fame is being the birthplace of Rubén Darío, the celebrated Nicaraguan poet. The learning center I'm visiting is managed by the international NGO Seeds of Learning, which supports education programs across Nicaragua. The center is an impressive complex, comprising a library, technology center, performance stage, dormitory rooms, and other facilities. Local students come to the center in the afternoon after school to do homework, take classes, and participate in other extracurricular activities. I see one group of young children struggling over a puzzle. I see another learning a dance on stage. Yet another group is practicing English with two Peace Corps volunteers. I'm surrounded by a whirlwind of activity I'd describe as enriching and fun.

Many of the students are from remote and poor villages. My understanding from speaking with the staff is that many people in the region (and most professionals) have simple phones, though a few have smartphones. Even having simple phones makes a big difference in their lives.

For example, at many of the more remote schools, teacher absenteeism has been a major challenge. In the past, students might walk for hours to get to school only to find that their teacher couldn't make it in. Now if the teacher has, for example, a sick child at home, she can call ahead to get a substitute. Alternatively, she can go to work but call home periodically to make sure her child is OK.

In Dario, even simple phones, which have proliferated only recently due to lower costs, represent a major step forward.

Whence the Internet?

Can we take a minute to remind ourselves about the remarkable appearance of the internet?

While the internet doesn't have an official start date or place of origin, most people agree it began with the computer networking efforts of the 1960s, much of it driven by the US Department of Defense ARPANET, which spurred the internetworking protocols and packet switching technologies that underpin the modern internet.

In the 1970s, the Internet Protocol Suite (transmission control protocol / internet protocol, or TCP/IP) led to increased networking in the US and Europe, particularly among supercomputing centers. By the late 1980s, the first commercial internet service providers (ISPs) appeared.

In 1989, Tim Berners-Lee, a researcher at CERN, the European Organization for Nuclear Research, developed the hypertext transfer protocol (HTTP), which allowed the linking of documents on different websites through "hyperlinks." As Berners-Lee humbly described the process years later, "I just had to take the hypertext idea and connect it to the Transmission Control Protocol and domain name system ideas and—ta-da!—the World Wide Web." If only all world-changing inventions were so easy.

The first website based on these protocols was launched at CERN in 1991 (and is still online—cool).

Mosaic, the first popular web browser, was released by the University of Illinois Urbana-Champaign in 1993. The brilliance of Mosaic is that it elegantly (for the time) wove together a number of internet technologies (FTP, NNTP, and others) and, importantly, allowed inline posting of photos and other images. It quickly turned the internet from a world of technical text to a world of real-life, beautiful, useful documents.

The team that developed Mosaic, led by Marc Andreessen, then created Netscape, a commercial version, in 1994. With the release of Netscape, more websites using the graphics capabilities of the new browser started to appear.

Later, in 1994, two Stanford engineering students, Jerry Yang and David Filo, created a rudimentary website called "Jerry and David's Guide to the World Wide Web." It became a popular reference, grew quickly, was renamed "Yahoo," and became the second prominent internet corporate success story (after Netscape).

Broadband started to arrive in advanced countries in the late 1990s, provided first by the phone companies (DSL) and cable providers. Feature phones got haltingly better during this time. Then Steve Jobs introduced the iPhone in June 2007. Many smartphone copycats appeared shortly thereafter, and the internet for many people moved quickly to their phones.

In developing countries, the first twenty years or so of this story never happened. Most countries skipped straight to mobile access, first with 2G mobile systems to allow calls and texting. In many countries, those systems are now upgraded to 3G.

So we're now at the stage, about twenty-five years in from the introduction of Mosaic, where about half the population of the planet has good internet access. The other half doesn't yet but will over the next several years.

Incidentally, I remember first seeing the internet in 1990. A friend of mine in grad school, Ian Freed, wanted to show me something cool. His desktop computer (with a forty-pound monitor the size and weight of a small refrigerator) had a modem that, through some unintelligible combination of beeps and hisses, linked to some distant system.

Some text started appearing (slowly) on his screen: "This is the University of Minnesota!" he said. "The service is called Gopher! You can read documents on their computer!"

I was happy he was so excited. But I can't say I shared his enthusiasm. I wasn't clear why I would want to read documents on a research computer in Minnesota.

Silly me.

I'm in the smart, clean, red-and-white offices of Claro, the Latin American mobile phone company (one of many controlled by Grupo Carso, the firm led by Mexican billionaire Carlos Slim Helú). I'm in southern Nicaragua in an appealing beach community called San Juan del Sur, which provides pretty vistas, an authentic Nicaraguan feel, and really good beachfront lattes.

San Juan del Sur has been a popular vacation destination for Nicaraguans for many years. Next door to the Claro office is the former vacation home of the Somoza family, the dynasty that controlled Nicaragua from 1936—when Anastasio Somoza García lured rebel leader Augusto César Sandino to peace talks and then murdered him soon afterward—to 1979, when the dynasty was overthrown by the eponymous Sandinistas. The house is now a hotel—Victorian and stately, if somewhat well-worn.

I'm speaking with Danilo, an employee sporting the same stylish red-and-white outfit as his office mates. On the walls are posters of hip millennials. Although Nicaragua is the second poorest country in the hemisphere (only Haiti is poorer), I feel like I could be in any mobile phone office in San Francisco.

I'm asking about short-term phone and data plans, since I'm in the country for just a few weeks. The best plan appears to be unlimited calls, 1 GB of data, and one hundred text messages for two weeks for about $6.50 (inclusive of taxes and everything else).

I also review the phones for sale. The cheapest smartphone is $103, manufactured by Huawei, the Chinese telecommunications giant. (Huawei phones have been dropped in the US by AT&T, Verizon, and others due to US government pressure to not sell Chinese phones that may be used to spy on Americans.) The cheapest feature phone, which allows calling and texting (but no internet), is $13. (I've also seen street vendors across the country selling smartphones and feature phones for about half these prices.) While many Nicaraguan professionals could easily afford these phones, they would be a stretch for the average school teacher, nurse, or policeman who makes around $200 per month.

I tell Danilo I'm interested in a plan. He asks for identification and accepts my California driver's license, which I find a bit odd. Would a Nicaraguan driver's license suffice in San Francisco?

I pay Danilo the equivalent of $6.50 for a two-week plan (data for Facebook and WhatsApp are free—paid for by Facebook). He pops a new SIM card in my phone, I answer a couple of questions on the screen, and I am done. The whole transaction takes maybe fifteen minutes—blazing fast by US standards.

Once I receive my Claro account, I start getting about a half dozen Spanish text "upsells" per day: "Today only—quintuple your data for a low price"; "Now, choose one friend for unlimited calling this week"; "Act now for a coupon to Pizza Hut."

I ask Nicaraguan friends about the upsells. They don't seem to mind them, mostly because many of them are genuinely valuable. The cell phone companies have become really good at understanding and selling to their clients. They've also become good at collecting data, since the phone is a data-collection dynamo. Most commercial businesses don't know much about you until you move fairly far up the economic scale. The credit bureaus in the US, for example, require bank accounts, loans, employment history, and other records before they have meaningful data. But phone companies are uniquely positioned to gather data and engage productively and immediately with even the poorest households. They know how much you spend and if you pay your bills. They know where you are. They know with whom you communicate. Poor people are typically economically invisible, but not to the phone companies.

With my Claro SIM card in my phone, I can surf the web from any place with coverage in Nicaragua, which includes all urban centers and many rural areas. I assume at some point I'll run through my credit and the surfing will cease. (Remarkably, it never does in two weeks.)

By US standards, the entire experience of dealing with Claro in Nicaragua is remarkably quick, easy, and cheap.

In many countries, the government subsidizes access to the internet. In the US, for example, the FCC maintains programs to subsidize poor or rural consumers, although funding on those programs is being cut back under the Trump administration.

While the global reach of the internet is increasing through wireless, satellite, and other technologies, local usage of the internet is often spurred by government programs. Regional efforts can target underserved communities and provide training and content development for new users.

In Mexico, for example, where internet access is defined as a civil right in the constitution, government programs provide free internet access. Mexico Conectado installs Wi-Fi hotspots in parks and public buildings in order to bridge the digital divide and provide better government services.

A similar government program in Colombia, Vive Digital, has promoted millions of new internet connections through the expansion of broadband and the distribution of computers in poor and rural communities.

In Australia, NBN has built a wholesale local-access broadband network with government support to serve disadvantaged communities.

Many countries have prepared forward-looking internet plans. A key component is often direct government subsidies for connectivity.

> I'm sitting in a park in San Juan del Sur, Nicaragua. My phone indicates a free hotspot called "Parque_Wifi." I link to it and find it painfully slow and barely functional. I see a few dozen folks on benches around me also looking at their smartphones, presumably competing for the same bandwidth.
>
> In 2014, the Nicaraguan government announced a program called "Park Wifi" to provide free internet access from every central park in Nicaragua. Nearly all cities in Central America are laid out around a central park. Even if the service is slow and unpredictable, the government views free (limited) internet as a reasonable "public good" for those that can't find better service at another hotspot or through their mobile carrier. It's a laudable goal.
>
> OK—I'm heading back to my hostel to use its internet, which is much better.

In order to promote broadband expansion, in 2010 the United Nations created the Broadband Commission for Digital Development (in 2015 renamed the Broadband Commission for Sustainable Development). The Broadband Commission seeks to promote global broadband for development, supporting the UN's 2015 Sustainable Development Goals, all seventeen of which benefit by wider adoption of broadband.

The Broadband Commission boasts an impressive roster of commissioners, co-chaired by Carlos Slim Helú, the Mexican telecom magnate, and Paul Kagame, president of Rwanda.

The commission issued seven targets for 2025, including nation-level policy plans, lower costs, broader penetration, and better services for banking, running small businesses, and so forth. Quantitatively, two goals stand out. First, the commission has set a target that in 2025 all countries will offer affordable broadband services, defined as less than 2 percent of monthly gross national income (GNI) per capita. Second, the commission seeks global broadband penetration of 75 percent. (Interestingly, this figure assumes penetration in least-developed countries of only 35 percent, a figure that the satellite companies would dispute as unambitiously low.)

In addition to the UN Broadband Commission, other organizations and research efforts track the global expansion of broadband. Tufts University, in cooperation with Mastercard, has developed a Digital Evolution Index (DEI), which ranks countries based on their progress in digital development. The DEI includes sixty countries evaluated across more than one hundred different indicators. In its main ranking, the DEI compares overall digital development to recent digital progress in order to see which countries are innovating and improving most quickly.

The 2017 DEI report identifies a number of "breakout" developing countries that are showing great progress in connectivity, including China, Bolivia, and Kenya. It also identifies some "watch out" developing countries where progress is halting, including Egypt, Pakistan, and Peru, and highlights the importance of focusing on mobile infrastructure by, in part, paying attention to government policies and initiatives to promote access, privacy policies, and security.

A separate group tracking global connectivity is Huawei, the Chinese networking company. Huawei releases an annual Global Connectivity Index (GCI), looking at forty indicators for fifty countries. The countries in the index represent 78 percent of the global population and 90 percent of the global gross domestic product (GDP).

The 2017 GCI report notes that while the metrics are climbing globally overall, there is wide divergence. Countries in the index tend to cluster into three groups: Frontrunners (average per-capita GDP

$50,000), Adopters (per-capita GDP $15,000), and Starters (per-capita GDP $3,000). The areas of focus for enhancing connectivity for the three groups vary widely.

The GCI suggests the gaps between the groups are widening. The rich, so to speak, are getting richer in both absolute and relative terms. This raises a number of policy-related issues so that, in the words of the report summary, the "digital divide doesn't become a digital chasm."

Finally, the World Economic Forum, working with the International Telecommunications Union (ITU), recently issued a working paper called "Connecting the Unconnected," which offers current data and program updates regarding global broadband. The document outlines the many reasons why 53 percent of the world's population (~3.9 billion people) are still offline:

- **Access.** One-third of the world's population is farther than one hundred kilometers from a fiber optic connection (although 84 percent do live within range of a 3G cellular network).
- **Infrastructure.** Many of the poorest on the planet don't have electricity—a prerequisite for internet access.
- **Cost.** Of the world's population, 57 percent can't afford internet access as offered.
- **Education.** Only 44 percent of the world's population have a secondary education or higher—a clear predictor of internet usage.
- **Relevance.** Many in the poorest countries don't see the relevance of online services (and may grow to be actively opposed).

The report describes that new technologies (satellites, balloons, drones) may help but are still several years from implementation. The issue of relevance and cultural acceptance may in fact prove to be especially daunting for the poorest regions. If people don't understand the internet or see its value, why bother? Will they even buy smartphones?

Quiz: Who sold the most phones in Africa in 2017?

a. Samsung
b. Apple
c. Xiaomi
d. Transsion

The answer is Transsion (who?!), with 38 percent market share in Africa and more than one hundred million smartphones sold in 2017. The Shenzhen-based company has focused on price (feature phones cost as little as ten dollars) and on customization for the African market. For example, many countries in Africa have multiple carriers with varying coverage. Transsion phones come with as many as four SIM card slots—a feature well appreciated in Africa. Transsion cameras also are designed to work well with darker skin and less light.

While Transsion brands are well recognized in Africa—Itel, Tecno, Infinix—they are still mostly invisible elsewhere. But that is changing fast. Transsion is expanding quickly in India. The company's research shows that Indians eat many meals with their hands and need their phones to work with oily fingers, so Transsion designed phones that do. Their phones are now the third largest seller in India (after Samsung and Xioami). The company also has launched products in Bangladesh, Pakistan, and Nepal.

Transsion smartphones typically retail for around one hundred dollars or less, but prices are dropping as the company scales production. The phones are currently manufactured in China, but new factories are being launched in India and Ethiopia.

The prices for smartphones are falling fast—and that is significant. But are the phones themselves any good?

I'm walking through Limbe, Malawi—the gritty, bustling industrial district adjacent to the regional capital, Blantyre. There are thousands of people on the sidewalks and in the streets, mostly men. The small, crowded shops are almost all industrial in nature—building supplies, agrochemicals, food packaging services. Several women sit on the sidewalk: one sells small mounds of peanuts, another sells bananas, a third sells what resemble deep-fried baseballs. My Lonely Planet guidebook, which is unfailingly cheerful about every city it describes, writes of this district, "You may have to change minibuses here, but it's best to head straight on." I can say with a high degree of certainty that I'm the only foreigner in Limbe this afternoon.

I enter a shop with a faded sign over the door reading "Sky Electronics." I'm in the market for a cell phone. The glass counters around the shop present maybe eighty different phone models, half feature phones and half smartphones. Everyone in the store is speaking Chichewa, but a helpful young man asks in English if he can assist me. I learn that the cheapest smartphone is something called A11 by Itel, one of the brands of the Chinese company Transsion. The A11 costs 35,000 Kwacha, or about US$47. I ask which phone is the most expensive, and the man tells me it's the iPhone X, at 1,000,000 Kwacha, or about US$1,300. I think I'll go with the A11.

The salesman unboxes the phone, drops in a battery, quickly goes through a few configuration screens in English, and then hands me the smart, lightweight device. It resembles my iPhone but is all black, shiny, and very comfortable in my hand. It's slick.

I ask if I can buy a SIM card and some data minutes. They don't sell those in the shop, so he asks the guard at the front door to run across the street, where there is a woman sitting at a table under a red Airtel umbrella. A SIM card costs 500 Kwacha (about 70 cents). I pay an extra 1,500 Kwacha (about $2) to load up on minutes and data. The guard returns in a bit with a SIM card and a couple of scratch-off data cards, and the salesman installs the SIM card, types in some numbers to enable the minutes, and hands me the device, all ready to go. The whole transaction takes about ten minutes.

Why again is this phone process so cumbersome and expensive in the US?

I assume there is a possibility the device will quickly fail, be painfully slow, or display other obvious shortcomings. When I get back to my room, I turn it on and log into the local Wi-Fi. I search Google. I pull up the *New York Times*. I play with the alarms and the calculator. I take a few pictures. I review the thirty or so apps that are conveniently loaded. I watch YouTube. I check Facebook (which has free data—paid for by Facebook).

My overall impression is that this thing is fast, powerful, and elegant. (And it's only $47?) While it isn't yet affordable to everyone on the planet, it almost is.

I feel like this thing I'm holding in my hand is a very, very big deal.

When users first get a smartphone, the device comes with certain apps preloaded. Typically front and center are two: Facebook, which is many new users' first stop, and WhatsApp (owned by Facebook), the dominant texting app in many countries

In many regions, Facebook also subsidizes data usage when users are on Facebook or WhatsApp; it is cheap or free.

Facebook cares a lot about increasing connectivity in developing countries. Its business depends on it. In 2013, Facebook grouped a number of efforts at expanding connectivity in developing countries into an initiative called Internet.org. The name confuses some, since it sounds like a neutral, nonprofit initiative when it is in fact a Facebook initiative. Current site branding refers to "Internet.org by Facebook."

When Facebook launched the initiative, much of the media attention focused on high-tech connectivity issues, in particular by comparing

Facebook's Project Aquila (which flies drones to provide internet access) to Google's Loon (which flies balloons). The more consequential Facebook program, however, is the Facebook Free Basics program, which provides free access to Facebook and other selected content in over sixty developing countries. The program is often controversial. On the one hand, Facebook is lauded for subsidizing access to Facebook and other resources. Much of the content Facebook includes is useful, including Wikipedia, Wikihow, and education and health sites. On the other hand, Facebook itself clearly benefits, as its application eclipses all others. Research in 2015 showed that 65 percent of Nigerians, 61 percent of Indonesians, and 58 percent of Indians agreed with the statement "Facebook is the internet." (Only 5 percent of Americans agreed.)

Controversies around Free Basics became particularly intense in India, to the point that Facebook canceled the service there in 2016.

After reaching its first billion users in 2013, Facebook's growth has been driven largely by developing countries. Only forty-one million of its second billion users are from the US and Canada. It is clear from the metrics—and anecdotally from traveling through developing countries—that Facebook is having remarkable success as the initial service for users coming online. And soon there will be three billion more users, a fact undoubtedly not lost on Facebook executives in Menlo Park.

I'm exploring rural Bali today with a local driver, a voluble and funny guy named Yanie, who seems to follow a certain pattern in leading tours.

First, we drive down rural, unmarked roads, sometimes hardly identifiable as roads at all. Next we stop and view something of wonderful beauty: a temple, a volcano, a waterfall, terraced rice paddies reaching to the horizon. Then we take selfies.

We repeat the process: Drive. Beauty. Selfies.

It surprises me that Yanie is happy to be in pictures and actually uses his phone to take lots of photos himself. Hasn't he been to each of these places countless times?

Then he explains: nearly all of his business is referred, and most of that comes through Facebook. He has found that keeping his Facebook feed current and compelling is really good for business. It's not much of an exaggeration to say that Facebook has become his business.

The same can be said for our surf instructor, who also gets referrals online. And the hostel where we are staying. And the little coffee shop with a small blue *f* logo at the counter. Often these individuals or organizations don't have web pages or other marketing. They do, however, have a Facebook page that is the centerpiece of much of their communications.

For many small businesses across Bali, Facebook is a godsend.

As broadband expands, more people get access to Facebook. But what other opportunities does internet extension make possible? Let me explore that question in a place characterized by past warfare, fierce animals, and amazing progress.

POSSIBILITY

I'm touring the E. O. Wilson Biodiversity Laboratory, named for the celebrated Harvard biologist. The facilities have what one would expect of a first-rate biology lab—DNA extraction equipment, refrigerated specimen storage rooms, and eager graduate students hard at work on plants and mammals and insects. There's even a pretty nice dining area. What is remarkable about the lab, however, is that it isn't located in Cambridge, Massachusetts. It's in Gorongosa National Park in Mozambique. Gorongosa, located in a remote section of the country, is both enormous in area—at nearly 1,500 square miles, it is larger than Rhode Island—and extraordinarily rich in biodiversity.

It is something of a miracle this laboratory is here at all. Gorongosa was once a showpiece of African national parks. But during the Mozambique civil war between 1975 and 1992, the rebels, called RENAMO, were based in and around the park. In more than seventeen years of conflict, nearly all the animals were killed (mostly to be eaten by hungry soldiers), and infrastructure was destroyed. In the 1990s, the park and surrounding communities were in a shambles.

Since then there has been a vigorous effort—led by the Mozambique government with considerable assistance from the Carr Foundation, an American philanthropic organization—to rehabilitate the park and surrounding settlements. Greg Carr is a technologist and successful entrepreneur whose first company, Boston Technology, is credited with popularizing voicemail.

In the 1990s, Carr became heavily involved in Africa, in part as a cofounder of Africa Online. He visited and fell in love with Gorongosa, and he saw the potential for the region—and all of Mozambique—in the park's rehabilitation. He also claims inspiration from Nelson Mandela, who told him that national parks needed to do more than just protect animals; they needed to become "human development engines."

Carr encouraged the government of Mozambique to create a long-term plan for the park, including strategies for wildlife restoration, environmental support, local community development, ecotourism, and jobs development. The government demonstrated its commitment, and in partnership, the Carr Foundation has pledged $90 million of support over thirty years.

Among many other initiatives, the Carr Foundation partnered with Professor E. O. Wilson to establish a research facility in Gorongosa, which Wilson has described as "ecologically the most diverse park in the world." Gorongosa staff built a small campus, shipped equipment, and—importantly—established good cell and internet access early on to enable global collaboration.

With this infrastructure in place, a team of Mozambican and international researchers collects and catalogs species (a number of which were previously unidentified), uploading descriptions, photos, sounds, and other information to online databases to be accessed by the global community of researchers. One Belgian researcher, Bart Wursten, showed me a plant that he discovered on Mt. Gorongosa and that is now named after him (*Impatiens wuerstenii*). A Mozambican researcher revealed a newly discovered species of flat gecko she proudly says is named after the park (*Afroedura gorongosa*). In the database, I saw gorgeous photos of beetles and heard chirps of bats and clicks of moths.

The overall story of Gorongosa is one of hard work, steady progress, and renaissance. The E. O. Wilson Biodiversity Laboratory, through its tight international collaborations that even a few years ago would have been technologically impossible, enriches both the Mozambicans who work there and the global community overall.

It is clear in traveling through rural areas of Africa, Asia, and Latin America that some new technologies are adopted everywhere quickly—others not so much.

Television, for example, immediately takes off. The poorest of families, even those lacking electricity, might somehow find an old TV, acquire a car battery, and jury-rig an antenna in order to grab a few channels. When that same family finds a bit more money, a satellite dish is likely to appear on the roof. People clearly like to be entertained and are willing to find the funds to allow it.

Simple feature phones also take off. People understand how they work and immediately see their value, and prices for both the phone and the service have dropped precipitously. Most of the people on the planet today can afford a simple phone, and most regions have at least spotty cellular coverage to allow voice calls.

Other technologies are making progress—but haltingly. For example, over a billion people still live without a lightbulb in their home. Currently small-scale solar—either individual lanterns or small systems with a couple of bulbs and a phone charger—have become high in quality and low in price. I see occasional solar lights or small solar panels in homes I visit, but they are far from prevalent. In some cases, governments are intervening. In India, for example, where over 250 million people have no access to electricity, the government is implementing an ambitious program to provide every unconnected home in the country with a "microgrid" (solar panel, battery, five LED lights, fan, and power plug) within a year. Prime Minister Modi wants the entire country to be electrified, even if at modest levels. Microgrids are limited, providing perhaps 125 watts of continuous power (which is much less than my home in California uses, even when all devices are turned off or in sleep mode). Having reliable light, a fan, and a cell charger, however, is a major step toward modernity and convenience.

I'm in a hardware store in Dario, Nicaragua. Actually "store" is an exaggeration—it's more like an overstuffed room half the size of a 7-Eleven, crowded with tools, plumbing fixtures, electrical supplies, and other knick-knacks I don't recognize. I ask the manager if he happens to carry solar lanterns. He appears with a compact Chinese fixture, about the size of a pint of milk, that sports a solar panel on the top. He demonstrates the device's three modes: "lantern" (for work), "flashlight" (for travel), and "disco" (for turning any hut into a party house)—seriously, there were flashing red, green, and blue LED lights on the top that could be turned on to make any situation seem festive. (What will the Chinese think of next?)

The lantern sells for about seven dollars, which strikes me as a bargain compared to what people in the region spend on candles and kerosene. I ask how many he sells. Not many, he says. People don't know about them yet.

It seems like only a matter of time before the lanterns become more common, but it hasn't happened in Dario quite yet.

If providing solar lights seems like an uphill struggle, replacing humankind's oldest technology—cooking fires—has proved to be nearly impossible in many places. In around a billion households across the planet, food is still cooked or warmed on fires built indoors. In the huts I visited in Malawi, for example, the preferred method is to place three rocks in the center of the living space, build a small fire, and push sticks into the flames from the sides. All the smoke plumes indoors. If there isn't enough wood, other materials will do (dung, crop waste, coal). The families I saw in Nicaragua living across from the dump burn whatever they can find there in order to prepare their food.

Aside from the environmental problems caused by cutting down trees and producing smoke, carbon dioxide, and many toxins, the health consequences of indoor air pollution are massive, leading, according to the World Health Organization, to *over four million* deaths per year. Indoor air pollution can reach levels that are *one hundred times* the acceptable level for particles and toxic substances. It also has a disproportionate impact on women and children, who spend more time near the fire. This is easily the biggest environmental health hazard we face as a planet, yet it gets relatively little attention. And technical progress to other forms of cooking has been slow.

Cheap, effective technologies exist for building ventilation or better fire containment, but for many reasons, they haven't been widely adopted. People don't recognize the health hazards. Smoke keeps bugs away, fires have always been used for cooking, and other technologies require funds to implement.

If households get a bit more money, they eventually make the transition to propane, which often isn't more expensive to use than firewood, although it does involve up-front costs for a stove and tank.

The household transition from feature phones to smartphones is common in countries I visit. The transition from wood fires to propane is not nearly as rapid. Government programs to promote propane exist, but they get complicated because they require large subsidies, which in turn can lead to theft and corruption. India has instituted a successful program for widespread subsidies for cooking fuel for the poor—reportedly the biggest cash transfer program in the world—and has benefited from tight controls on access to the subsidies. But this success story hasn't been frequently replicated elsewhere.

In summary, some technologies need little help for adoption, such as television or feature phones. Some important advances require government assistance, such as clean cooking technologies. But where does broadband fall on this spectrum? On the one hand, in areas with sufficient population and wealth density, broadband is quickly provided by the private sector and is widely adopted. On the other hand, in areas with insufficient population or wealth (in both developed and developing countries), broadband requires government assistance to promote access and adoption. This "mixed story" is one reason several billion people still lack access to the internet. In many environments, while access is expanding, it still requires a push.

Humankind has used fire for cooking, heat, and light for millennia. We've also used signal fires for communications—sending messages from community to community through flashes of flame.

Ironically, we are still using light for communications—but we've gotten much better at it. While we might think of the internet as being carried by electric cables, the internet is principally light. We have built a global nervous system of fiber optic cable, with information cascading around the planet in threads of glass, animated by the blinking of lasers.

On Fiber and Light

Can we take a minute to celebrate the marvel of the fiber optic cable?

Scientists have known for 150 years that glass can be used to guide light. By the 1980s, manufacturers were able to make highly transparent threads of glass the width of a human hair and over a hundred kilometers long. Simultaneously, laser technology was getting cheaper and smaller, and digital data processing was getting faster. These technologies converged to give us fiber optic cables.

Why is fiber such an improvement over copper cable, which was first laid across the Atlantic in 1858? For starters, fiber optic cables have an astonishing capacity for carrying information. A single fiber can carry three million simultaneous phone conversations. Since a cable can comprise over one thousand fibers, this means a single cable could support three billion conversations—or half the planet speaking with the other half simultaneously.

Furthermore, light travels efficiently with very low attenuation. Signals can maintain sufficient strength for over one hundred kilometers before needing a boost. Fiber optic cables carrying information with pulses of light aren't subject to electromagnetic interference the way typical copper cables are. The signals avoid corruption. Eavesdropping is much more difficult because light doesn't give off electromagnetic "signals" the way electricity does.

But my favorite characteristic of fiber optic is its main ingredient in a thread: silica (a.k.a. sand). While copper cables around the world are highly prone to theft (copper can cost a few dollars a pound, and large cables will weigh tons), their fiber optic counterparts are not. If thieves want silica, it's a lot easier to pilfer the beach. (Unfortunately, that sometimes happens: a few years ago an entire beach in Jamaica disappeared overnight due to thieves stealing hundreds of tons of sand.) In India, when companies lay fiber optic cable, they typically leave a few coils laying around the worksite so potential thieves can see for themselves that the materials in the cables aren't valuable.

When powerful technologies converge—in this case, materials science, laser technology, electronics design, data processing, and complex manufacturing—magic can happen.

Fiber optic cable proliferated throughout the developed world at the end of the twentieth century. The first transatlantic fiber optic cable, called TAT-8 (for Transatlantic cable #8) was constructed in 1988. A rapid proliferation of higher-capacity cables followed.

However, the first fiber optic connection to the African continent only arrived in 2000 with the SEA-ME-WE3 cable, which stretches from Germany, through the Red Sea, to India, Southeast Asia, and Australia. That cable connected to Egypt and Djibouti.

Meaningful connections to Africa didn't appear for another decade. Since 2010, however, every year has seen logarithmic growth in capacity. Current capacity to all the countries of East Africa is approximately twenty-four terabits per second (Tbs) over multiple cables, a figure soon expected to grow to nearly ninety terabits due to the completion of a major new cable (DARE). West Africa's capacity is approximately forty-five Tbs, a figure soon to expand to nearly two hundred Tbs due to the completion of three new major cables (SAIL, SACS, EllaLink).

New cables not only introduce capacity; they also introduce redundancy. Undersea cables are periodically damaged unintentionally, such as in the commercial shipping incident that caused an outage in all of Somalia. With a new web of connections, outages will be less prolonged and severe.

At the same time that undersea cables are proliferating, hundreds of projects are laying cable across the continent itself. Liquid Telecom, operator of the largest fiber optic network across Africa, has laid over 50,000 kilometers of cable. In 2017, Google laid about 1,000 kilometers of fiber in Uganda and is currently laying 1,000 kilometers more in Ghana. Facebook also plans to add nearly 1,000 kilometers of fiber in Uganda.

With added capacity comes added competition—and lower prices. Nic Rudnick, the chief executive of Liquid Telecom, estimates that the price of moving a megabit of data from London to Lagos has dropped over several years from $600 to $2.

Fiber optic cable represents only part of the broadband solution in Africa, since it will never reach many rural areas due to cost. It is serving, however, more and more urban areas, providing fast and reliable access to businesses and consumers across the continent.

It also isn't the only technology expanding quickly in Africa and other developing countries. Many other behind-the-scenes technologies are playing a consequential role. One such technology, viewed as critical by network engineers, is the proliferation of internet exchange points (IXPs) in developing countries. An IXP serves as a country-level or regional gateway between different networks, obviating the need to send traffic to distant regions or countries in order to be routed correctly. In other words, IXPs provide local shortcuts for internet traffic, which greatly lowers cost and latency. According to the UN's 2017 State of Broadband report, "According to Packet Clearing House, 24 more countries established a new IXP over the twelve months between mid-2016 and mid-2017 (of which eleven were African). By mid-2017, 119 ITU Member States now have IXPs, compared with 76 ITU Member States which do not. The total number of IXPs in ITU Member States globally is 471."

The African countries recently adding IXPs were Benin, Botswana, Burkina Faso, Côte d'Ivoire, Republic of Congo, Madagascar, Malawi, Mozambique, Rwanda, Sudan, and Zimbabwe. This increases the total number of African countries with IXPs to 29. The EU has 145 IXPs, and the US and Canada have 84. But now even Malawi has one—another step in The Great Connecting.

I'm sitting in the courtyard of a primary school in Nmunda, Malawi, located in a very poor corner of the country near the Mozambique border. This isn't, however, just any school courtyard. It is part of a full campus the size of a small village, with an impressive brick wall enclosing a dozen classrooms, a library, administrative buildings, a soccer pitch, a parking lot, teacher housing, service buildings, an acacia forest, and a grove of fruit trees. The school, only a few years old and serving over six hundred local children, is here because of the cooperation of the Malawi Ministry of Education with the Hamels Foundation, an American charity started by Cole Hamels, a star major league baseball player, and his wife, Heidi.

The arrival today of Heidi and a group of girlfriends from the States has been greeted with unbounded exuberance by the kids.

Several of the foreign guests are working down a line of children, offering high fives and hugs. Another woman is leading a group of small children in a round of "Heads, Shoulders, Knees, and Toes." Another is surrounded by a cluster of teenage boys singing a Justin Bieber song.

The school, an inspirational site for the local community, stands in stark contrast to the school down the road, which has about one hundred desks for over seven hundred students and needs to conduct two-thirds of its classes outside under the trees.

I'm here today in part to explore the role the internet has played in the school (bandwidth was an early investment) and the opportunities the school may have to link to other programs around the world. It's all quite miraculous to contemplate this remote school collaborating with other schools and organizations around the planet in meaningful ways.

I'm mostly thinking, however, about this crazy commitment by foreigners from far away to try to help the kids here in Malawi. Cole and Heidi have donated generously—I've heard millions of dollars—to support rural Malawi. Most people will never know about this. It's inspirational. There is a lot of love in the air this morning.

The staff and teachers are giving speeches describing their programs and expressing their gratitude for international support. The kids are singing songs and reciting poetry. There is a great deal of understandable appreciation for the Hamels Foundation, and I guarantee there is also considerable bewilderment. Did that young woman of the group of visitors pay for all this? How is that possible? And what is this thing called baseball? (By the way, before marrying Cole Hamels, Heidi achieved her own degree of celebrity as a much-followed contestant on season six of *Survivor*—good luck explaining that to the Malawians.)

I'm here to see the school's internet connections, but it's clear that what is really important—what has come first—are the human connections that make it all possible.

The Hamels School has broadband because it could afford it. That means that the students, who represent many hundreds of households, will have access to the internet. Others from the community will also benefit from the Wi-Fi.

I see this dynamic in many settings. Even if most people can't afford broadband (yet), the fact that it exists in some places, even if it is driven initially by foreigners, is in itself a game changer.

The GAIA office down the road from the Hamels School has Wi-Fi costing a few hundred dollars per month. I've seen higher-end hotels in many countries somehow figure out how to get Wi-Fi. My language instructor in Nicaragua, who lives in a small home and makes a modest income, installed an expensive microwave receiver on her house to access Wi-Fi so that she could teach Spanish by Skype to former students in the US, which easily pays for the connection.

What this means is that for certain regions, relatively low penetration figures can mask the true impact of broadband. Even a little bit of broadband getting to a region for the first time can bring many benefits.

I'm in a Seeds of Learning community learning center in the outskirts of Tipitapa, Nicaragua. It is bustling with around fifty kids playing games, assembling puzzles, reading books, and doing art projects. In the back of the room, a couple of teenagers read through Wikipedia pages on laptops for their homework assignment.

Behind the learning center is an eighty-foot radio mast with a small microwave antenna on the top. Two years ago, an international donor paid for an internet connection for the learning center. In this part of the country, that requires building a large tower.

The center now has free Wi-Fi. What has that meant to this poor community? The residents are still trying to figure out the value. It's now simple to get information. It's easy to find news. It's effortless to find Facebook (as best I can tell in the countries I visit, Facebook finds *you*).

It's often not so easy to quickly locate those resources that are genuinely useful. But with time, they start to appear.

One teacher proudly showed me a number of crafts she had completed with the kids, with ideas she found online. I heard in detail about teachers coming to the center from around the region to download curricula and videos to their phones to use in the classroom. Apparently, one source of great didactic materials for teachers in the region is Pinterest. Others are experimenting with online education involving agricultural programs.

The woman running the learning center told me that laborers learned recently that they can come to the center to get a printout of their police registration document, which is necessary when applying for a new job. This saves a trip into the capital (which takes a couple hours each way), a long wait in line, and about seventy cents of bus fare. They find the convenience of strolling to the learning center to get their document absolutely terrific—a miracle, really.

And every day the local residents discover something new. There is a lot to explore when a smartphone meets Wi-Fi. Recently the internet service unexpectedly was out for about a week. Many people were very frustrated. Laborers could not believe they had to take a bus to Managua to get their police documents.

The Role of Broadband in Development

International global development agencies spend a lot of time debating and prioritizing development efforts. Those priorities play a major role in shaping national and regional agendas.

The most prominent global development prioritization efforts are led by the United Nations. In 2000, for example, the UN established the Millennium Development Goals (MDGs), a set of eight development targets to be achieved by 2015. Each goal had associated metrics and timelines.

Progress toward the goals was uneven at both country and international levels. Some goals were not achieved (such as reductions in child and maternal mortality rates), while other goals were achieved early (such as global reduction in poverty, mostly thanks to China and India).

In addition to meeting with mixed success, the goals prompted debate in international development circles about whether the best, most legitimate eight goals were chosen. There was a parallel debate around the chosen success metrics.

Despite the shortcomings and disputes, the MDGs are widely credited with increasing attention, funding, and coordination around fundamentally important global milestones.

As the end of the fifteen-year window approached, the United Nations launched a follow-up fifteen-year effort entitled the Sustainable Development Goals (SDGs). Careful to avoid criticism for a hasty selection of targets, the UN considered literally hundreds of possible goals, eventually winnowing the list down to seventeen—including 169 "targets" and 304 "indicators." (The large number of goals, targets, and indicators unleashed a new wave of criticism.)

So what does all this have to do with broadband expansion? A lot, actually.

It's easy to point to the key role broadband will play in achieving each of the goals. For example, goal 1, "elimination of poverty," will be directly impacted by economic growth. Goal 2, "zero hunger," will be directly impacted by better communications, coordination, and policy implementation.

Some goals stand out as particularly tied to broadband access. Goal 3, "good health and well-being," and goal 4, "quality education," are closely intertwined with the internet, given the large number of online services for both health and education. Goal 8, "decent work and economic growth," is as well.

> The UN has laid out an ambitious set of goals to achieve by 2030, and broadband will play a major role.

As several billion people come online in the next few years, those of us in developed countries need to remind ourselves that our own experience of coming online was very different than theirs will be. Technologies in the past were simpler. We also had a big head start in terms of understanding technology, media, and information systems. Today, in contrast, much more powerful technologies are appearing in much less developed environments. When I came online in the early 1990s

- the services were very limited—accessed through a snail-paced 2,400 baud modem. Today's service is lightning fast in comparison.
- there wasn't much to see. I remember having a sense of basically everything being online—which wasn't much at that point. Today, online information feels infinite.
- services were a supplement (information, education, health, government services) to my everyday life. Today, these are entirely new services for many poor, rural users.
- I had a strong understanding of the underlying technology and a familiarity with technology issues in general. Today, most new users likely have no idea about how it all works.
- I understood that the computer could help supplement (or replace) my bank, my library, my school, my health clinic, and other organizations. Today, most new users don't have a bank, library, school, or health clinic to replace.

- I had a set of skills to help navigate and interpret the new media. Today, most newcomers are media newbies, with limited exposure even to TV or radio.
- there wasn't much concern about malicious content. Today, we have sophisticated fraud, identity theft, fake news, and other online hazards.

Many good things will come from new populations getting access to broadband. But let's not kid ourselves: it's complicated. It reminds me of the joke about our gradual adjustment to the internet:

- *Internet 1998:* Don't meet people online; don't get into strangers' cars.
- *Internet 2016:* Summon strangers from online; get into their cars.

We had decades to habituate to the current online environment. Newcomers will have only minutes.

I've watched people in developing countries be handed their first smartphone. They have absolutely no idea of its power. How could they? In developed countries, we understand (and marvel at) the fact that our phones replace dozens of devices we used to carry around: a phone, camera, music player, calendar, address book, flashlight, radio, tape recorder, alarm clock, calculator, photo album, video player, and much more. People in developing countries frequently have never owned (or seen) any of those things, so they have no way to understand even a fraction of the capabilities of the new device.

Even more challenging, many of the worst aspects of online life—fake news, fraud, porn—will likely find them quite easily.

Shouldn't we be doing all we can to make sure their first online experiences are as useful as possible? What they need is a "welcome wagon." *Collins English Dictionary* defines a welcome wagon as a "welcoming service that provides information about a community to new residents."

Currently, most people are handed a smartphone with some preloaded apps (determined by both the manufacturer and the service provider) and maybe some literature about their service plan. In some parts of the world, this is supplemented with information about Facebook Basics. That's about it.

What would an ideal new online experience be for a new smartphone user?

- a welcome video in an appropriate language from a credible person to greet the new user
- a quick tutorial on five really useful basic services (e.g., phone, texting, camera, weather information, calendar)
- a curated list of one or two outstanding sites for news, health, education, finance, and government services
- what to do if you need help
- simple tutorials for more information, including how to avoid problems with fake news, fraud, or other issues

Ideally, this would be developed and managed by either governmental or international organizations to provide credibility and avoid favoring any given corporation. The Wikimedia Foundation is probably the best example of a global organization that has figured out the governance issues, quality control, and sustainability challenges of providing information for all. Could a group like that build and curate digital welcome wagons in every language on the planet?

If you pay attention, you'll see that The Great Connecting is already happening in consequential ways almost everywhere. For example, today I'm observing door-to-door outreach for HIV testing by GAIA in rural Malawi. I'm with Dr. Jay Levy from the University of California, San Francisco. Levy, the codiscoverer of the HIV virus, is an academic celebrity in the US. Here in Malawi, his celebrity is elevated to rock-star status—academic celebrities don't come to Malawi. (Celebrities don't come to Malawi in general—with the exception of Madonna, looking to adopt.) Levy and his wife, Sharon, wanted to see the very front lines in the battle against AIDS, and to their credit, here they are.

Because Levy is here, so are the local and international press. I'm chatting with Frank Phiri, the regional Reuters correspondent. As we bounce over dirt roads in our four-wheel-drive vehicle in the rural south of the country, Phiri is pounding out a story on his laptop. He also is editing photos and video he has shot. He calls his editor on his smartphone, assuring her his story is en route. And in the next hour or so, he knows we will be on a stretch of road within range of a local cell tower (he is familiar with coverage and dead spots across the country better than anybody), and he'll be able to upload his stories and images to be passed on to the international wires. The upload won't be easy or cheap—he probably subscribes to all three mobile services in the country to offer redundancy and find the strongest signal. But at the end of the day, Phiri is posting stories, nearly in real time, from the most remote corners of the country. Until recently, that was unthinkable.

Broadband is now reaching remarkable places for the first time around the planet. For The Great Connecting truly to reach its full potential, however, it requires investment on the part of countless organizations.

Let's get a glimpse of one company's investment in a place about as far away from southern Malawi as one can get.

INVESTMENT

I'm standing in the middle of the Nevada desert near the town of Winnemucca, a place most travelers avoid. The largest highway in the state, Highway 80, passes north of town, and few travelers stray beyond a gas station or diner along the frontage road.

They may be missing out. Winnemucca is home to the Buckaroo Hall of Fame (don't use the word *cowboy* in these parts). The downtown includes several well-worn casinos, brothels (legal), and a bank once robbed by Butch Cassidy's gang. Until recently, the world's largest potato dehydration plant was located nearby. Winnemucca proudly refers to itself as "The City of Paved Streets."

Through the years, the town has boasted significant Latino, Chinese, and—surprisingly—Basque communities. Today's residents are mostly employed by several large gold mines in the area.

I'm actually a few miles out of town at the Winnemucca Municipal Airport. It's a sleepy place, with several parked Cessnas, a few helicopters, and incongruously, two large seaplanes. The desert stretches in every direction to distant, imposing mountain ranges. A jackrabbit nibbling on shrubs pays me little attention.

Behind me, however, is something striking: a massive, red metal cube, about fifty feet on a side. As you drive up in your car, you can't help but think, *What is that thing? An amusement ride? A jumpy house? A new-age shrine?*

Up close, you see it is actually a very complex structure: a large and imposing crane and gantry system with winches and cabling. Wind socks and antennas line the top. On the crossbeam are printed the words "Big Bird."

This system, custom built by Google's Project X (now simply "X"), is actually the company's giant launch system for Loon, the internet balloon service. The balloons themselves are enormous—almost fifty feet across—and when filled with helium, they are exceedingly difficult to control in the wind during launch. So Google built a custom system that can stabilize the balloon, block the wind, and automate the launch process. In the past, it would take sixteen employees to launch one balloon, with timing dictated by the weather. With this giant contraption, it now takes only four employees, launching balloons at will.

This morning, the team appears to be preparing for launches later this week. They are raising and lowering the jib crane. They are testing the large hangar doors used to block the wind. They raise and lower a large perch used when releasing the balloon. I see enormous tires on the unit, which allow the team to pivot the contraption to best block the wind.

The complex where the launcher is located has pickups and forklifts and shipping containers, all suggesting a lot of activity at this remote "balloon port." Satellite dishes dot the premises.

X has big ambitions for Loon, a technology to provide internet access to the three billion people currently offline. That sounds crazy—but the Nevada desert is a really good place for crazy. South of here, for example, tourists would gather in Las Vegas in the 1950s to watch aboveground nuclear explosions while hotels would run beauty contests for "Miss Atomic Energy." West of here, tens of thousands of people gather each Labor Day for Burning Man, a pop-up utopia. Along State Route 375, a remote highway in the east of the state that runs next to the top-secret Area 51, there were enough reports and sightings of strange events that the state officially renamed the road the "Extraterrestrial Highway."

Now Google is trying to launch balloons from Nevada to provide broadband coverage for the planet. That is far from the craziest idea to be explored in this area. But I'm just wondering, *Giant balloons? The women of Mutambi? Really?*

Google wants everyone on the planet to be online (and, while at it, to use Google services). It understands, however, that traditional fiber optic plus cellular connectivity isn't going to work for the half of the planet still without broadband. So Google has been investigating other approaches for expanding connectivity.

Loon (which is short for *balloon*) aspires to provide internet access to rural and remote areas by using high-altitude balloons. The balloons, residing in the stratosphere—about eighteen kilometers up—form an aerial wireless network providing coverage similar to that provided by cell towers. The oblong balloons, filled with helium, are about fifty feet across and forty feet high. They are made of a translucent silver material and carry, dangling about eighty feet below, a payload of communications equipment, solar panels, and batteries.

The promise of using balloons in the stratosphere to provide cellular coverage is that it will be much cheaper than building cell towers. One balloon can provide the coverage of dozens of cell towers. With lower costs for coverage, it becomes economically viable for telecommunications firms to serve low-density or poor populations across the planet.

The balloons navigate by changing their altitudes to find winds blowing in appropriate directions. They include an inner envelope filled with helium and an outer envelope that can ingest or expel air as ballast to change altitude. (Loon staff named the fan unit that manages this after the *Saturday Night Live* character "Franz" because it wants to "pump you up.")

Initially, Loon had assumed the balloons would simply ride the major wind currents eastward, launching a "necklace" of balloons to circumnavigate the planet. With more experience, however, and with wind mapping and forecasts provided by the National Oceanic and Atmospheric Administration (NOAA) and other international agencies, Loon discovered they could find winds blowing in appropriate directions to navigate wherever they wanted (or stay more or less stationary) just by adjusting the elevation of a balloon. The company now manages navigation and placement of the balloons through complex data analysis and artificial intelligence technologies, requiring relatively few balloons to provide extensive coverage.

The initiative, active since 2008, has tested balloons in six continents, logging over twenty-six million kilometers around the world. Many balloons stay aloft for over one hundred days, with the record being nearly two hundred days.

Most recently, Loon was credited with providing over two hundred thousand Puerto Ricans with free internet services following widespread outages on the island. After Hurricane Maria, Google moved quickly to deploy balloons, requiring the prompt cooperation of many organizations, including the government of Puerto Rico, the FCC, FAA, FEMA, AT&T, T-Mobile, SES Networks, Liberty Cablevision, and others. Communications is a team sport.

Google engineers determined that having five to seven balloons over Puerto Rico at any one time was sufficient to provide full coverage for the island, based on complex simulations of weather patterns. Google decided to launch balloons from its Winnemucca test site, where launch facilities were ready to go. Apparently it isn't too hard to get a balloon from Winnemucca to San Juan (3,409 miles as the balloon flies). Google quickly established cell coverage across the island.

Prior to disaster relief in Puerto Rico, Loon provided emergency internet coverage in Peru following flooding in 2017.

I'm walking through a generic office parking lot in Mountain View, California. The building in front of me is drab brown and unremarkable.

To one side is a peculiar row of employee vehicles, all electric and from seemingly every manufacturer—VW, BMW, Audi, Smart, Honda, Nissan, Fiat, Kia, Chevy, Tesla. I've never seen such an array of purely electric cars. Is this the future? What is this place? I don't see signage anywhere.

I'm looking for X, the secretive Google research and development lab. I figure I must be close: as I was parking my car, two autonomous automobiles with the word *Waymo* on the side patiently waited for me to finish before they headed off on their merry, robotic way.

Earlier today, I happened to be at Google's main headquarters, dubbed the "Googleplex," which is brightly colored, architecturally compelling, and full of signs and symbols that make it clear where you are, including cute statues of the green Android mascot, "Bugdroid."

Here at X, there is simply a parking lot and an unremarkable brown building.

The mundane appearance, however, belies what is inside. X is Google's (or now technically Google's parent company, Alphabet's) self-described "moonshot factory," whose mission is to invent and deploy breakthrough technologies that live at the intersection of a big problem and a radical solution.

It's not known publically how many projects X is currently working on. What is known is that a number of projects have "graduated" to become stand-alone companies. The best known of these is probably Waymo, the self-driving car company whose vehicles patiently waited for me to navigate to my parking place. Waymo recently placed an order for twenty thousand Jaguar electric cars to be built with Waymo technology and serve future self-driving services.

Other companies that have graduated from X include Wing, a drone delivery service; Verily, a life-sciences firm; and Chronicle, a cybersecurity company. More than a dozen companies overall have spun out of X. One company under development of potential relevance to The Great Connecting is Free Space Optics, which is trying to replace fiber optic cable with laser relay stations that beam data through the air.

X also publicizes projects they investigate and decide to cancel, including studies of a personal jetpack (found to be too loud), hoverboard (found to have too few societal benefits), and teleportation (found to violate the laws of physics).

Inside the lobby of X, which has the open loft feel of a company with plenty of resources, things are more interesting. A video on the wall highlights a number of companies that have spun out of X as well as various projects under way. In the rafters of the lobby hang prototypes from different initiatives, including some sort of drone aircraft and a large translucent balloon.

I'm here to meet with Scott Coriell, head of global communications for Loon. In addition to working in various communications roles in technology and government, Coriell also lived for two years in Malawi, working for a health nonprofit based in Lilongwe. So my first question cuts to the chase: When will rural Malawi get broadband?

Coriell relates that we may know the answer to that relatively soon. Loon recently signed its first commercial client, Telkom Kenya, to provide 4G internet access to rural parts of the country currently without broadband. Loon expects the network to launch in 2019 and is currently navigating technical and regulatory details with Telkom Kenya, the government of Kenya, and international bodies. The hoops Loon needs to jump through are many: permission from the International Civil Aviation Organization (ICAO) to fly balloons, spectrum requirements in Kenya, airspace permission in Kenya, and rights to land its balloons in the country when required. In some cases, laws (or even jurisdictions) may not yet exist for flying internet balloons in the stratosphere, so Loon needs to work with appropriate Kenyan authorities to authorize its activities.

When Loon is operational, users in Kenya probably won't know anything about it. They simply will see a signal on their cell phones from Telkom Kenya and will access calling and data services as they would anywhere else in the country that has traditional cell tower coverage. They probably won't ever even see the balloons, which, when flying in the stratosphere, look like tiny silver specks.

Loon has a lot to learn in Kenya. Does the system work? Is there sufficient demand for bandwidth in rural areas to make the economics viable? Can the electronics dangling below the balloon handle the bandwidth requirements? Are balloons placed over more urban areas also a promising line of business?

If Loon is successful in Kenya, many other telecom companies around the globe are in line to be future partners (and those discussions have begun). Loon plans to launch balloons bound for Kenya from its new Puerto Rico launch facility. It can launch a balloon in about thirty minutes. With sufficient demand, dozens of launches per day are possible, hundreds per week. Scaling up to other countries or regions should be relatively straightforward.

Is there a future for thousands of Loon balloons providing bandwidth around the globe? Will the balloons ever reach Mutambi? Loon is doing its best to make this happen—and as quickly as possible.

Other companies are tackling the broadband challenge with different technologies. Facebook, for example, shares Google's goal of connecting the next three billion users and has experimented with giant solar-powered drones to provide bandwidth to rural and remote regions around the world. The drones, part of Facebook's Project Aquila, fly at sixty thousand feet (about the same as Loon balloons) and are aloft for months at a time.

An Aquila drone was tested for the first time in June 2016. It flew well—before crashing. Mark Zuckerberg lauded the first flight and said that Facebook learned a lot, but he warned that Project Aquila still faced engineering challenges related to weight, power, control, speed, altitude, load, and communications. This stuff is hard.

Facebook has also been using the Aquila platform to research more efficient radio technologies to increase Aquila's bandwidth capabilities and, potentially, to be used by satellites.

The vision for Aquila had been to have a fleet of drones communicating among themselves by laser and staying aloft for months at a time (while trying to avoid Google's balloons). Recently, however, Facebook announced it will stop manufacturing drones and leave aircraft development to the aerospace firms now entering the space.

Airbus, for example, recently set a new flight-endurance record of twenty-six days with its solar-powered drone called Zephyr S. The drone charges its batteries by day, flying at an altitude of about sixty thousand feet. At night, batteries power two small electric motors as the drone makes a slow descent to about fifty thousand feet. By day, the drone climbs back to sixty thousand feet. Airbus will soon be launching a larger drone, called Zephyr T, to be used for bigger payloads, including communications equipment to provide broadband.

Other companies have their own plans for extending broadband networks in novel ways. The Airborne Wireless Network, for example, proposes to outfit hundreds, eventually thousands, of commercial aircraft with telecommunications equipment capable of communicating with both ground stations and other aircraft. These "minisatellites" would form a mesh network serving worldwide data and communications providers. The firm has tested initial prototypes using two aircraft and plans a twenty-aircraft test shortly. The global rollout is scheduled for 2021.

Another more terrestrial example of expanding coverage is happening in the Australian outback. Seventy percent of Australia—a big, sparsely populated country—currently lacks cell coverage. But even in remote areas, there are lots of Toyota Land Cruisers crisscrossing the terrain. Flinders University, along with Toyota and Saatchi & Saatchi Australia, has proposed outfitting Land Cruisers with communications hubs capable of "store and forward" messaging. Each "mobile hotspot" would include Wi-Fi, ultrahigh frequency (UHF), and mesh networking capabilities with a range of 25 kilometers. Messages would be passed from vehicle to vehicle until reaching an internet-connected base station. The LandCruiser Emergency Network wouldn't provide true broadband, but it would offer messaging services useful for communities with few other options.

The internet is reaching the most remarkable places.

I'm in the jungle in Mozambique and am eye-to-eye with a pack of wild dogs. I'm not nervous, however, for two reasons. First, the dogs have just eaten—the picked-through remains of a waterbuck are nearby—so they are more interested in rolling in the grass than pursuing me. Second, and more consequentially, they are behind an imposing-looking fence, as they are currently in a holding facility in Gorongosa National Park.

Wild dogs once ran throughout this area but, like most of the other animals, were decimated by the Mozambique civil war. Now they are being reintroduced to the park. This pack was airlifted from South Africa. (Amazingly, my guide, who worked previously in South Africa, recognizes one of the animals, an alpha male named Hercules.)

Soon the pack will be released into the park, which I have to imagine is wild-dog heaven: nearly 1,500 square miles of wilderness teeming with impalas and other treats, with few other competing predators around for now.

When they are released, each animal will wear a tracking collar to assist the research scientists in understanding how the dogs move and adapt. Because cell coverage is not available in much of the park, the tracking devices communicate by satellite. Communications of this sort have revolutionized research biology. Because of improvements in satellite technology, devices are getting smaller and cheaper. These days, small devices are even placed on birds (apparently the Internet of Things includes the Internet of Wings).

Next up for Gorongosa will be the reintroduction of leopards. Actually, only female leopards will be reintroduced, at which point the males somehow will magically show up, migrating from distant territories. The leopard progress and romance also will be tracked every step via new and improving satellite technology.

The Carr Foundation provides financing and technical support for the reintroduction of species in Gorongosa. Because of his clear dedication to the park and the obvious progress that has been made over the past decade, Greg Carr has become something of a household name in Mozambique. I hear him described by park staff and local residents with a combination of respect, appreciation, and bafflement. They wonder why an American would dedicate so much personal money and time and passion over decades to help the country of Mozambique. Why a thirty-year commitment? Why not demand more personal credit?

It's impressive. I'm here to explore The Great Connecting, but I am witnessing the human connecting happening simultaneously thanks to people like Greg Carr.

Google may have balloons and Facebook may have drones, but the technology that is likely to revolutionize broadband extension into rural areas is satellite, specifically a new generation of small, low-earth-orbit (LEO) satellites currently being developed and deployed by a number of companies.

There are about eight hundred communications satellites in orbit today. Most are in geosynchronous orbit, about thirty-six thousand kilometers above the equator. At this altitude, a satellite orbits the earth in twenty-four hours, appearing to remain stationary over one point. That simplifies communications because antennas can continuously point at one spot in the sky. Geosynchronous communications satellites, however, are very expensive to design, build, test, launch, and deploy, costing at times more than $1 billion. Time from design to deployment can be many years. Satellites can be as large as a small bus. Because geosynchronous communications satellites are so far away, a signal takes over half a second to get there and back—an eternity in today's hyperconnected world.

A number of companies are rethinking the approach to communications satellites. Since the costs of electronics and launches are coming down, what if instead of placing a few dozen really big, expensive satellites into geosynchronous orbit, we could place thousands of small satellites into orbit very close to earth?

The most consequential company exploring new approaches is SpaceX. SpaceX has plans for its new Starlink constellation to include nearly twelve thousand satellites. These will operate close to earth—between 550 and 1,325 kilometers in altitude. Prototype satellites are already in orbit, with operational satellite launches beginning in 2019 and the rest rolling out over about five years. Limited service from the initial eight hundred satellites would be available globally in 2020 or 2021. The FCC license requires that half of the satellite constellation be launched within six years and the rest within three years after that.

In early 2018, SpaceX launched two test Starlink satellites, called Microsat-2a and Microsat-2b. The satellites, in an orbit of about 315 miles, are testing optical communications between each other as well as K_u / K_a band communications (the preferred bandwidth for SpaceX) with ground stations in Redmond, Washington (where the initiative is based); SpaceX headquarters in Hawthorne, California; Tesla headquarters in Fremont, California; and other sites, including three specially configured mobile vans.

The satellites reportedly weigh four hundred kilograms each, or about one-twentieth the weight of a typical commercial communications satellite in geosynchronous orbit. SpaceX's Falcon 9 has a payload capacity to low earth orbit of about 23,000 kilograms, suggesting that the company could launch perhaps fifty satellites of this size at a time. (They haven't verified this, but one has to assume that they are working hard to get satellite size down and numbers per launch as high as possible.) SpaceX already launches ten Iridium satellites (which are much larger) at a time. India has launched over one hundred microsatellites (satellites weighing less than 50 kilograms) on one rocket. SpaceX also plans to sell launch services to other satellite networks using the same satellite bus and requirements for its own Starlink network.

Low-earth-orbit satellites can be much smaller and offer internet latency periods of twenty-five to thirty-five milliseconds, equivalent to (or better than) many cable and DSL systems. Because Starlink satellites will be much closer to earth, the coverage area of any given satellite will be relatively small—a circle with about a 1,000 kilometers radius—requiring a large number of satellites in the network. The service will principally serve individuals and small businesses using laptop-sized antennas. SpaceX is promising "fiber speeds."

SpaceX plans to launch the full complement of nearly twelve thousand satellites by 2025. Internal SpaceX documents estimate that in 2025, its launch business might represent $5 billion in annual revenue, while global communications services might represent $30 billion in annual revenue (based on forty million subscribers). The global launch business is much smaller than the global telecommunications business—by more than an order of magnitude—so SpaceX is motivated to expand into telecommunications as a lucrative, adjacent market.

SpaceX estimates satellite lifespans of five to seven years before starting deorbiting procedures, which suggests that launches of new satellites will be required indefinitely. SpaceX reportedly has over five hundred staff working on the system, the majority of whom are in its Redmond, Washington, office, where the Starlink network is being designed and built. Previously posted job descriptions include almost every task in satellite design, which suggests that SpaceX is bringing everything in-house. This in turn suggests that the company believes there is a significant opportunity for innovation; otherwise, it would rely more heavily on established players. SpaceX recently leased a new, large research lab space near its Redmond office for testing new satellite technologies.

Gwynne Shotwell, president of SpaceX, commented after the initial launch of the prototype satellites, "Although we still have much to do with this complex undertaking, this is an important step toward SpaceX building a next-generation satellite network that can link the globe with reliable and affordable broadband service, especially reaching those who are not yet connected."

SpaceX is planning to build thousands of inexpensive satellites. They are also working on technologies to lower the costs of launch in order to make Starlink affordable.

I'm walking along a boat channel in San Pedro Harbor, south of Los Angeles. On each side of the channel are rusted train tracks and dilapidated warehouses. Few people are around on this hazy afternoon. The place feels a bit like the setting for an LA noir detective movie. I'm half expecting to see a drug deal in the shadows or somebody disposing of a body.

Several vessels are moored in the channel: a tugboat, a large ship that appears to be a research vessel of some sort, and some smaller craft. Across the water, however, is a boat that stands out. It's a new vessel that's painted a crisp blue, gray, and white and sports the sleek lines of a Coast Guard cutter. On the back half of the vessel are four enormous poles rising up and out over the water. Between the poles is enough room for an enormous square net of perhaps fifty meters on a side.

The boat, called *Mr. Steven*, is playing its own role in The Great Connecting. The craft is operated by SpaceX, and its role is to head out to sea to try to catch the payload fairings from rocket launches from California.

Fairings are the giant clamshell-like structures at the top of a rocket that protect the payload. When the rocket reaches orbit, the fairings release and fall away, typically burning up in reentry. SpaceX, however, is trying to reuse as much of each rocket as possible. It has mastered returning the first stage safely, having now landed over twenty-five rockets either back on land or on a barge at sea. Now it is working on capturing the fairings by having each fall back through the atmosphere (slowing from Mach 8), deploy a parafoil, and be guided carefully back to a designated landing spot—in this case, the enormous net on the back of *Mr. Steven*.

It seems like a really hard task. (It also seems a bit far-fetched—really, that might work?). But each set of fairings costs about $6 million, so it is a task worth tackling. As Elon Musk points out, if there were a pallet of $6 million of cash falling toward the ocean, wouldn't you try to catch it?

Mr. Steven just got back to port last night. Two days ago, while somewhere out in the Pacific, it just missed catching a fairing. SpaceX has now tried three times, with three near misses. This week it recovered both fairings intact from the ocean. I see them on the back of the boat's deck. Once the fairings hit the seawater, however, they can't be reused because it is too difficult to sterilize them to the levels required by ultraclean payloads. So SpaceX will keep trying with its giant net. It has recently announced a plan to quadruple its size. It all sounds nuts.

I've seen a blooper reel of all the rockets SpaceX crashed before figuring out how to land them correctly. People scoffed at that idea once too. Now every rocket executive around the planet at a company not called SpaceX is very concerned.

If (when?) SpaceX succeeds in catching fairings, the cost of launching rockets will drop by about another 10 percent, which drives down the price for the full industry, which makes the price of communications satellites lower, which provides more economical options for serving the billions of people who don't currently have bandwidth. That boat over there doesn't seem like it should have much in common with a hut in Malawi, but the two are actually closely linked through The Great Connecting.

SpaceX isn't alone in the race to satellite-based broadband; it isn't even leading the race. OneWeb, which received FCC permission before SpaceX, plans to launch as many as 1,980 satellites and aims for full global broadband service by 2021. LeoSat plans for 108 satellites. Iridium is deploying 66 next-generation satellites (with the help of SpaceX launches).

One firm has already implemented satellite broadband using a small and growing network of medium-earth-orbit (MEO) satellites. O3b Networks (now part of SES) currently maintains sixteen satellites at an altitude of 8,000 kilometers, which is about a quarter the distance of its geosynchronous competitors. Additional satellites are scheduled to be launched in 2019.

The O3b network provides backhaul services to mobile providers as 4G subscribers grow, according to the company, from 1.6 to 3.8 billion by 2020. The network also serves multiple niche markets, such as emergency response and cruise ships.

The name O3b, by the way, stands for "other three billion"—a reference to those on the planet currently without broadband.

Viasat, a satellite communications firm that has been providing services for years (with current satellite technology that is slow and expensive by terrestrial standards), claims it will be the first firm offering true global broadband access. The company is launching three high-capacity geosynchronous satellites between 2019 and 2021, each with network capacity comparable to the total of "the approximately 400 commercial communications satellites in space today." Viasat has a nice corporate tagline about linking the next three billion: "So everyone can have a voice. Imagine that."

Finally, a lot of eyes are watching Facebook to see what its next moves are around satellite communications. Facebook has been interested in satellite technology and can leverage its work to date on its Aquila drone project. Facebook's first satellite, designed to support broadband in rural Africa, unfortunately blew up on the SpaceX launchpad in August 2016.

Now Facebook has reportedly registered a new subsidiary to build low-earth-orbit (LEO) satellites. The subsidiary, called PointView Tech, plans to launch a demonstration satellite in 2019 to investigate using the E-band spectrum for communications. E-band promises much higher data connection speeds than those planned by rivals, but it needs to overcome some challenges, including absorption by rain and other particles. E-band is also used by the Facebook Aquila drones, so the company has experience with the technology.

The PointView Tech initiative puts Facebook in direct competition with SpaceX. There doesn't appear to be much love lost between Mark Zuckerberg and Elon Musk. They have engaged in a public feud around the future of AI and its role in society. Musk recently deleted all Tesla accounts from Facebook. SpaceX also mistakenly blew up Zuckerberg's satellite, which also probably didn't help relations.

So a lot of serious players are building satellite constellations, and most of the work is being driven by SpaceX.

The Role of Steerable Antennas

A new generation of low-earth-orbit satellites will rely on many new technologies: miniaturization, power management, solar power, battery technology, laser communications, and many others. One required technology to allow all this to work involves steerable antennas.

Historically, communication satellites have often been placed in geosynchronous orbit because of one great advantage: satellites in geosynchronous orbit appear stationary in the sky. Satellite dishes or antennas tracking the satellite don't need to move. Nongeosynchronous orbits require antennas to move to track the satellite, which can add a lot to the complexity and cost of the antenna. Precision motors are required to move the device up and down or side to side to track satellites. For decades, the Soviets employed the "Molniya orbit," which required satellite dishes to nod up and down from the horizon.

As companies contemplate placing thousands of satellites into low earth orbit and all the advantages that kind of deployment confers (less latency, smaller satellites, lower cost), a major challenge appears: How do you design an antenna to track satellites and navigate frequent handoffs from one satellite to another? And if the antenna is moving in a plane or car, how does that factor in?

Fortunately, there has been great progress in a new generation of "steerable antennas," also known as "phased array antennas." Researchers have essentially built the steering elements, until now managed through motors, onto a chip. Flat-panel antennas are being designed that can track satellites and manage frequent handoffs from one satellite to another.

The technology is proven, and a number of agreements have been signed between antenna technology firms and satellite companies. The antennas are technically sophisticated and currently very expensive, but with millions likely to be purchased by businesses and consumers for broadband access, companies are optimistic that prices will fall to a few hundred dollars each, making Wi-Fi hotspots around the globe economically viable. Soon, you won't be able to escape Wi-Fi.

I'm at the North Pole, and my Wi-Fi isn't working.

Well, technically I am seven miles above the North Pole, flying from the Middle East to California. I'm in a double-decker airliner carrying over five hundred passengers on a fifteen-hour flight. It's negative forty degrees outside, and I'm almost directly over the North Pole—so maybe it's understandable that the Wi-Fi isn't working. Remarkably, for much of the flight, it worked just fine. The aircraft is communicating with satellites, but apparently satellite coverage at the North Pole still isn't great. Yet.

SpaceX is the current dominant force in the launch game. As satellite technology evolves, however, so simultaneously do other launch capabilities, driven by some very entrepreneurial efforts, all seeking to drive down costs.

For example, Stratolaunch, a start-up aerospace firm founded by Paul Allen, is pioneering a new mobile launch system that would lower costs for deploying small satellites to orbit. Stratolaunch has introduced an enormous, twin-fuselage aircraft designed to carry up to three rockets that can then launch satellites into LEO. The aircraft, with the largest wingspan of any plane ever flown (150 feet greater than a 747), would fly to thirty thousand feet before firing the rockets. At that point, the aircraft would return to its staging area, refuel, reload, and be ready for three more rockets. This approach in theory will be much faster, cheaper, and more flexible than launching rockets from the ground.

It is estimated that the Stratolaunch aircraft will be able to launch a payload of five thousand to ten thousand pounds to low earth orbit, which is around a tenth of the payload of a Falcon 9 launch. But this method should be much cheaper and more flexible for microsatellites at low orbits.

Paul Allen passed away in October 2018. Work on the Stratolaunch aircraft nonetheless continues, aiming for first flight in 2019.

Another Seattle entrepreneur is also developing new launch capabilities. Jeff Bezos's rocket start-up, Blue Origin, is reportedly making steady progress in rocket and propulsion design. Blue Origin, like SpaceX, is focused on designing reusable launch systems in order to keep costs low. The company's current reusable rocket, called New Glenn, has already lined up paying customers (OneWeb, Eutelsat, and others) for launches beginning in 2022. It maintains major facilities in Washington, Texas, Florida, and Alabama and is funded at $1 billion per year from Bezos's Amazon stock sales.

Virgin Orbit, an initiative funded by Richard Branson, also aspires to launch satellites. LauncherOne, a two-stage air-launched rocket with a payload of around 300 kilograms, will be carried to high altitude by Cosmic Girl, a 747 previously operated by Virgin Atlantic.

In the world of launch and satellite services, Elon Musk, Mark Zuckerberg, Paul Allen, Jeff Bezos, and Richard Branson have all funded serious initiatives in a somewhat odd billionaires' race. Given the money and technology (and egos?) involved, we're sure to see rapid progress on many fronts.

Thankfully, nonbillionaires are also hard at work. Rocket Lab, an American aerospace firm, plans to use its Electron rocket to launch payloads of 150 kilograms into orbit. Using launch facilities on New Zealand's North Island, Rocket Lab aims for dozens of launches per year at a budget price of under $6 million per launch. It has conducted successful test launches of its two-stage rocket, and it plans commercial launches of CubeSats soon. One way that Rocket Lab keeps costs low is by manufacturing its Rutherford rocket engine principally through 3-D printing.

What's a CubeSat?

CubeSats are miniaturized satellites that comply with internationally agreed-upon standards, including component cube dimensions of 10 centimeters on a side and less than 1.3 kilograms of weight per unit. Imagine a square container with a liter of water—that is about the size and weight of a CubeSat.

Because they are so small and primarily use commercial, off-the-shelf components (mostly designed for cell phones), CubeSats are fast and cheap to design and deploy. Historically, they have been launched as secondary payloads on larger launches. Over 800 CubeSats have been deployed to date, and at least 1,200 more are planned for orbit. A new industry of launch services specifically targeting CubeSats (and other small satellites) is taking shape.

The simplicity and low costs of CubeSats mean many groups can now become involved in space science. Universities, high schools, and individuals have all designed and launched CubeSats. Some have even been funded by Kickstarter campaigns.

Developing countries are also getting in the game. For example, Kenya recently designed the CubeSat 1KUNS-PF, which was carried to the International Space Station by a SpaceX resupply mission and from there launched into orbit. Over the course of eighteen months, it will assist with mapping Kenya, monitoring the coastline, and spotting illegal logging. To date, eighty countries have launched CubeSats.

Another start-up, SpinLaunch, is following a novel strategy to launch CubeSats and other small satellites into orbit. It is reportedly building a giant catapult that will spin a rocket in a circle before flinging it toward space at hypersonic speeds of up to 5,000 mph. At that point, the rocket will ignite to carry the payload into orbit at 17,000 mph. The advantage of this method is that the rocket's energy initially would come from electricity on the ground rather than from rocket fuel. The cost savings would translate into much lower launch costs of about $500,000 per satellite. The catapult would also be designed to function many times per day. The idea sounds nutty, but the firm is viewed as credible, having raised $40 million from Google, Airbus, and venture capital firm Kleiner Perkins.

Many other companies are getting into the CubeSat game. Sky and Space Global, a company based in London, plans to launch two hundred CubeSats into low earth orbit in order to provide telecommunications services in Africa, Latin America, and elsewhere. The satellites will be deployed in near-equatorial planes, reaching fifteen degrees north and south of the equator.

Satellites will communicate with ground antennas that provide Wi-Fi hotspots or potentially with a new generation of twenty-dollar Android phones capable of direct communication with the satellites.

Sky and Space Global aims to build and launch the entire constellation of two hundred satellites for $200 million, a fraction of the cost of even one geosynchronous communications satellite. The firm has contracted with Virgin Orbit for launch services.

I'm in Hawthorne, California, walking down Crenshaw Boulevard. Ahead of me, towering over the intersection and nearby buildings, is a SpaceX Falcon 9 rocket.

The rocket, positioned next to SpaceX headquarters, is massive. Despite being only the first stage, it reaches perhaps twelve stories into the sky. I'm particularly struck by the four legs of the rocket, which hold it upright. They are enormous, reaching two stories up the body of the rocket, each the diameter of a large oak. They look much more spindly in YouTube videos.

These specific legs are actually a bit famous. This rocket (Falcon 9 booster B1019—which is too much information, but the booster does have its own Wikipedia page) was the first to return to earth and land successfully after deploying communications satellites into orbit.

SpaceX has ambitions to build a far larger rocket. Initially called the BFR (Musk would only say that *B* was for "Big" and *R* was for "Rocket"), the next-generation launch system is now called "Super Heavy." It will likely be big enough to fling hundreds of Starlink satellites into orbit with each launch.

The private sector isn't alone in investing in The Great Connecting. Many national governments are also involved globally.

The Chinese government's Hongyun Project, for example, plans to launch three hundred satellites into low earth orbit, with the network operational in 2022 and complete by 2025.

Other Chinese firms, typically with close coordination with the government, are also involved. The Chinese firm LinkSure Network has announced plans for a constellation of 272 satellites with aspirations to provide free Wi-Fi to regions currently without coverage.

Many other countries are also involved. For example, in early 2018, SpaceX launched Bangabandhu-1, a communications satellite for the Bangladesh Telecommunication Regulatory Commission. The $250 million satellite will provide broadcasting and telecommunications services to rural areas in Bangladesh from a geostationary orbit. The Russians recently launched the communications satellite Blagovest 12L to provide, among other things, telecommunications and mobile broadband services to rural areas in Russia.

What about Space Debris?

Won't the launching of thousands of satellites into orbit create a big space junk problem? The answer is "maybe." We need to be careful.

The good news is that space is big, even in near-earth orbit. There is a lot of room for a lot of satellites. If you think about the number of boats the oceans can accommodate and then realize that space is much, much larger, you can appreciate that there is a lot of room to work with.

All new satellites need to have launch approvals and also decommissioning plans (typically involving falling back into the atmosphere and burning up). SpaceX and OneWeb, for example, have committed to one-year decommissioning plans for satellites at the ends of their lives. SpaceX has also asked for FCC permission to deploy a subset of Starlink satellites at only 550 kilometers, which would make decommissioning much easier.

Governments are pretty good at tracking larger pieces of space junk and identifying potential problems. The International Space Station is periodically moved in order to minimize the chances of collision with a piece of space junk. Around twenty thousand man-made objects are currently tracked in space (although they need to be big enough to track—the United States Strategic Command estimates many millions of smaller items are also in orbit).

One concern is that a single collision can lead to the creation of much more space junk, which in turn could collide into other satellites. Computer models show that a chain reaction of this sort is possible, known as the "Kessler syndrome." This isn't hypothetical: at least five satellite collisions have resulted in increased space debris. Both the International Space Station and the Mir Space Station have sustained damage from collisions with space debris. Space crowding is particularly acute at the poles: many satellites maintain polar orbits in order to have complete coverage of the earth, which means the orbits all cross at the poles.

Some scientists argue that we are already in the early stages of the Kessler syndrome, with collisions certain to increase in coming years. Other scientists argue the problem is under control. But everyone agrees we need to be careful about how we plan for and manage all future objects we launch into orbit.

The private sector and governments are playing a major role in The Great Connecting, but the nonprofit sector also is making a major investment.

As I walk through the hallways of the most influential nonprofit organization in the world, I'm thinking about how much of an impact someone could have if he or she could spend $10,000 per minute.

The Bill and Melinda Gates Foundation, the largest charitable foundation on the planet, makes nearly $5 billion in grants annually (nearly $50 billion total to date). Its headquarters' campus in Seattle is pristine and lovely, characterized by expansive glass, soaring rooflines, and lovely views. The simple architecture contrasts sharply with the Frank Gehry–designed Museum of Pop Culture across the street, funded by the other Microsoft founder, Paul Allen.

That building is raucous and colorful. But the contrasts don't stop with architecture. Downstairs at the Gates Foundation is a "Discovery Center" where visitors can learn about vaccine strategies and mosquito eradication. Allen's museum celebrates rock 'n' roll. One display in the Gates Foundation Discovery Center is a high-tech toilet with an interactive guide to how the "urine path" is separate from the "fecal path." (There's also a sign reading, "Interested in a demo? Our staff will be happy to show you.") At the Museum of Pop Culture across the street, you can record your own DJ track. There aren't many visitors this morning at the Discovery Center, which is unfortunate. The displays are genuinely important, and it is laudable that the foundation is working hard to publicize these topics.

I'm meeting today with Matthew Bohan, a senior program officer at the Gates Foundation who works in the program providing "financial services for the poor." Bohan's and his colleagues' efforts support government and private-sector partners in providing online banking services for the world's poorest. Foundation grants aim to strengthen the payments infrastructure, increase the use of financial services, and improve policy and regulatory efforts.

It's tough work rife with obstacles. For example, many people in poor countries who sign up for online banking services don't have government-issued identification. Other forms of verification, such as voice biometrics, are required. Even simple security systems such as a four-digit PIN don't work if the population is innumerate. People can't remember four digits if they can't even count. Remembering a PIN for them is like an average American remembering four random symbols in Hindi—unlikely to work.

Despite the challenges, many countries are making great progress with online banking. In just the last six years, 1.2 billion people have gained access to bank and mobile money accounts. That number is climbing quickly.

The Gates Foundation doesn't directly support the extension of broadband into resource-poor environments, but once broadband arrives, a long list of foundation programs can immediately take place. Financial services for the poor support online banking. Foundation vaccine programs rely heavily on geographic information system (GIS) services. Global health programs rely on real-time data analysis. Foundation-supported researchers in developing countries collaborate online with colleagues across the planet.

Overall, the expansion of broadband greatly opens the playing field for Gates Foundation programs. Once a community is online, billions of dollars of philanthropic initiatives become possible.

The Reach of the Bill and Melinda Gates Foundation

The Bill and Melinda Gates Foundation is the largest charitable foundation in the United States, with total assets of over $50 billion. That figure, however, greatly understates the Gates Foundation's true financial clout.

First, the total doesn't include the commitment made by Warren Buffett in 2006 to donate over 70 percent of his fortune to the Gates Foundation. Buffett's net worth is currently listed at over $80 billion, despite the fact that he's already made multibillion-dollar annual grants to the Gates Foundation since making his initial pledge.

The total assets of the Gates Foundation also don't include the net worth of Bill and Melinda, currently listed at over $90 billion. The Gates have said that most of their fortune will eventually be given to charity, chiefly to the Gates Foundation.

So the Gates Foundation assets are better calculated not as $50 billion but more realistically as over $150 billion, or ten times the size of the Ford Foundation, the next largest charitable foundation in the US.

Even this understates the financial influence of the Gates Foundation. It is a "spend-down" foundation, with the mandate to spend all assets and go out of business within twenty years of the deaths of Bill or Melinda, whichever is later.

Most foundations spend 5 percent of their assets each year, the minimum required by law. Depending on the performance of their investments, this generally means that endowments increase in size over time, with foundations lasting indefinitely. Spend-down foundations, in contrast, have higher annual giving because they are not aspiring to last forever. This makes a big difference. Depending on investment performance, a spend-down foundation may grant triple the amount annually of a typical foundation of comparable size.

The Gates Foundation spends most of its funds in three areas: Global Health, Global Development, and Education. These were the priorities identified by the Gateses early on, and the foundation has been very consistent in sticking to these priorities.

Anyone working in a field touched by Gates Foundation philanthropy will describe the outsize role the foundation plays in the sector. For example, the foundation has been very involved with vaccine research and vaccine implementation globally and totally dominates efforts in this important realm.

The Gateses and the Buffetts could be spending their vast fortunes on anything they want. But they choose to invest in programs directly helping many of the neediest on the planet. Thank goodness they do.

Across the planet, private companies, governments, and nonprofits are making huge investments in the technologies that lead to The Great Connecting:

- fiber optic cable
- cell phone towers
- new devices
- satellite technology
- launch services
- internet infrastructure
- software and mobile apps

How much of an investment? It's hard to say—probably a number with eleven zeros (hundreds of billions of dollars annually) is a reasonable estimate. But are investments paying off? Is this relationship showing signs of promise? To find out, I'll need to visit a place far from Seattle where everyone I meet is excited about the growth and opportunities of expanding broadband.

GROWTH

I'm walking through the halls of a large conference center that could be in any city in the US—except for the dudes dancing, beating drums, and wearing leopard skins. The Information and Communications Technologies for Development Conference (ICT4D) has drawn eight hundred participants from eighty countries to the Mulungushi International Conference Centre in Lusaka, Zambia. Everyone is here for one reason: to explore and demonstrate how internet technologies can benefit the poorest people on the planet.

It's hard not to be impressed—and optimistic. A quick walk through the exhibit hall finds one group showing off a tablet-based app to track vaccination compliance. The next group is enabling mobile payments for government services. The next guides mosquito (malaria) eradication efforts through geographic information systems (GIS) technologies. Nearby I see an agricultural app to help farmers with fertilizer use, a mobile phone survey company, and a smart system to support health workers in the field. There are education groups, refugee support organizations, telecommunications initiatives, and even a technology incubator with the hip Zambian name of BongoHive. My favorite presentation of the event has been about using mobile apps to support toilet sales in rural Africa or, as the presentation was titled, "Using 0s and 1s in the Cloud to Combat #2 in the Open."

I'm also happy to see some big players here: Mastercard, SAP, Salesforce, Motorola, Barclays, and Facebook all have a presence. Apparently these companies believe the bottom three billion people on the planet represent a market opportunity.

I'm literally surrounded by hundreds of organizations looking to reach the poorest households on the planet with useful services—as soon as those households get broadband.

It is tempting to talk in detail about the many innovative online services currently available to those in developing countries with access to broadband (I never knew that smart drones, driven through internet connectivity, could cheaply and selectively apply fertilizer based on infrared imaging, for example). Here are just a few cases of influential services in different sectors already making an enormous difference in poor communities around the world.

EDUCATION

BYJU'S is a popular online education service in India providing online tutoring for students in grades four to twelve. The firm offers an education app, free for thirty days, that has been downloaded more than fourteen million times. BYJU'S uses innovative teaching tools, such as a video of a Bollywood dance troupe moving rhythmically to demonstrate the Pythagorean Theorem. If students don't have a sufficient internet connection to download materials, BYJU'S will send them a memory card with appropriate resources. More than nine hundred thousand Indians have signed up for the paid service.

BYJU'S lists prominent investors including Sequoia Capital and Tencent and a valuation of $800 million, and it joins a long list of online education initiatives reaching large numbers of students in developing countries. Top universities in the US and Europe offer many courses for free. For-fee services like Coursera or EdX boast tens of millions of users taking thousands of courses. Private schools, such as Bridge International Academies, rely heavily on online services. Khan Academy, the reigning king of online education, boasts over one hundred million annual users in dozens of topics across more than twenty languages.

India faces profound challenges in education: more than half of fifth graders, for example, can't read at a second-grade level. Online services such as BYJU'S aim to help.

BYJU'S is one of many online education services targeting developing countries. In order to test which services are best, the XPRIZE Foundation has launched a $15 million Global Learning XPRIZE to be awarded in 2019 to the team with the most effective open-source software solution for teaching reading and mathematics skills to children in developing countries.

IDENTITY SERVICES

Aadhaar is the Indian government's resident identification number system, which has registered about 1.2 billion Indians—nearly everyone in the country. While Aadhaar doesn't directly relate to internet communications, it both depends on and enables many online services.

Aadhaar issues every registered user a twelve-digit national identification number that links to biometric data—typically a photo, ten fingerprints, and two iris scans. Soon it will include facial recognition services as well.

Registration with Aadhaar is voluntary, but so many government and commercial services (welfare programs, pensions, banking services, mobile phone accounts) are now linked to Aadhaar that almost every adult in India uses the service.

Like any powerful technology, Aadhaar brings both profound advantages and significant risks. An Aadhaar identification number allows Indian citizens—including the poorest—ready access to information and services heretofore unavailable to them. It increases efficiency and decreases corruption. For example, India has subsidized programs to help fund cooking fuel so that households don't need to rely on burning wood or coal indoors. It makes it possible to guarantee that the right people get the fuel while minimizing fraud. On the other hand, centralized identity services such as Aadhaar can lead to data breaches, fraud, and abuse. There is a vigorous debate in India about Aadhaar's proper role and how to make it more secure and effective.

Many countries are developing or refining their own identification systems. Aadhaar is currently the largest and, arguably, most influential personal identification system on the planet, but other innovative systems are starting to appear.

Back at the conference center, I'm listening to a presentation about Element, a new biometric software service headquartered in New York. Element wants to create an identity service that works in poor areas using only smartphones, not requiring specialized equipment (like Aadhaar). The company also want to avoid the privacy issues and stigma of facial recognition software.

Element takes a picture of a person's palms, analyzes the photo, and encodes the analysis. With that information, the system can verify identity with high precision. The company is currently experimenting with an infant identity service that takes a photo of the bottom of a baby's foot.

It is possible that in the future, you will only need to show your hand to be verified by Element software to sign up for a government program, register for school, or just buy a Coke.

LOGISTICS

Zipline is a California-based company that combines sophisticated drone technology with expanded communications coverage in rural areas of Africa in order to deliver lifesaving medical supplies. From a central staging area, Zipline flies medical supplies by drone, on demand, to rural clinics within eighty kilometers of the hub.

In 2016, Zipline launched its initial service in partnership with the government of Rwanda. Now distant clinics in hard-to-reach areas (of which there are many in hilly Rwanda) can use the Zipline app to send a request for blood, medicine, vaccines, surgical supplies, or other pressing needs. A drone is loaded and launched, carrying a payload of up to 1.5 kilograms. When the drone arrives at the clinic, it circles a predetermined retrieval area, sends a text that it has arrived, and drops the payload by parachute. The drone, powered by electricity, returns to base for recharging and its next assignment.

Centralizing the storage of key medical supplies, now possible because of Zipline, allows for lower inventories, better security, and safer management of medications (often requiring refrigeration).

Since operations began in Rwanda, Zipline has completed more than 1,400 flights covering 100,000 kilometers. It has announced a major expansion into Tanzania, including four delivery centers supporting 2,000 flights per day to more than one thousand clinics across the country.

HEALTH

MomConnect in South Africa is a phone-based text support system for pregnant women and young moms. When pregnant women in South Africa visit a government health clinic, they are automatically referred to the service. About 70 percent of pregnant women sign up nationwide to MomConnect, which sends text messages to their phones concerning health, nutrition, and parenting advice from pregnancy through the first two years of their children's lives. Each of the nine hundred thousand active users (at any one time) receives messages tailored to her stage in pregnancy or childcare, her language (all eleven official languages of South Africa are supported), her age (very young moms have specific needs), and her HIV status. Users receive messages a few times per week (about 250 total messages over 2.5 years).

Users can ask questions about pregnancy or childcare. The MomConnect help desk, staffed by health professionals, receives about one thousand questions daily. Users can also provide feedback about the quality of service they are receiving at local clinics and submit any suggestions or complaints, allowing the National Health Service to identify and address problems quickly.

The mothers get tailored information, can ask questions, and can provide feedback. For women with otherwise limited access to health care, this is unprecedented. It also presumably makes the women feel highly valued.

One challenge that MomConnect faces is that texting fees, all paid by the service, are expensive (currently over $100,000 per month). The company is currently conducting a trial with WhatsApp aimed at the subset of moms with smartphones. Already MomConnect sees that moms use WhatsApp four times as much as they use text. WhatsApp fees will likely be lower, although final details are still being determined.

With the support of the Gates Foundation, MomConnect has also established a Mobile Engagement Lab to conduct controlled experiments across the network in order to learn and improve. Which messages are most effective? How tailored can communications be? What results can be demonstrated? MomConnect is already reaching hundreds of thousands of women in South Africa and is designed to be increasingly useful as it grows.

I'm in the Banani District of Dhaka, Bangladesh, feeling lost. I've looked for District Block A, which I located; then Road 27, which I think I found; then Building 47, which I believe I'm in; then the ninth floor, which is where I hope I am standing. Now I need to find apartment 9B. Am I even close?

I then spot a sign on one of the doors to my left that reads, "mPower Social Enterprises."

When the door opens, I no longer feel lost. The suite of rooms is full of software developers and product managers huddling over laptops and scribbling plans on glass partitions. They are too busy fixing bugs in their code to pay much attention to me. I see a shelf displaying various awards, much like the awards in my company's lobby in Washington, DC. The only striking difference between this suite of offices and others where I have spent many years is that the front conference room doesn't have any furniture. Meeting participants kick off their sandals at the door and sit on the ground with their laptops. This strikes me as a brilliant way to keep meetings short.

mPower is a Bangladeshi company started by international development professionals who saw the potential of applying information and communications technologies (ICT) across their country. Community health workers have historically used paper forms when going from house to house. Using mPower technology, they can now carry a tablet or smartphone to facilitate their work. This changes everything. They can gather more information. They can compile the information daily or in real time. The information can be used immediately for smart decision support. Systems can provide health workers with better questions or advice. Data are backed up. Citizens have new channels for providing feedback.

The company has collaborated with many international partners and funders in developing an open "smart register platform" called openSRP that NGOs in developing countries can use. Increasingly, government health systems themselves are adopting the technology.

Other mPower programs involve technology assistance for both veterinary and agricultural extension service workers. One current technology focus is on making apps smarter by incorporating more detailed weather, satellite, and soil data into the existing GIS capabilities. That way recommendations can be customized by region—or by individual farmer or plot of land.

Once field staff have smartphones, a world of opportunities opens. Software designers around the world, including these in this hidden technology refuge in Dhaka, are chasing those opportunities.

FINANCIAL SERVICES

GiveDirectly is a New York–based nonprofit that provides unconditional (i.e., no-strings-attached) cash transfers to people in extreme poverty through online banking technology.

Anyone who has worked in global development has been frustrated at times by inefficiencies. For example, a taxpayer in the US gives money to the federal government, which gives money to USAID, which gives money to an international consulting firm, which gives money to a regional nonprofit, which arranges contracts to build and manage a hospital, which serves poor people. Might it be better simply to skip all the intermediaries and give money directly to poor people?

What happens when you give money directly to the poor? How do they spend it? Is it wasted? Is it better to give money to everyone in a community or just the poorest? Women or men? Fewer installments or more? Shorter commitments or longer? How do outcomes compare with traditional aid programs?

GiveDirectly has explored these questions through well-designed randomized controlled trials reviewed by third parties. The conclusion so far? While unconditional cash transfers aren't without challenges, they are highly effective and economically efficient at moving people out of extreme poverty. They are apparently more effective than most aid programs and should be used as a yardstick against which to measure the performance of other development efforts, much as a stock index fund can be used as a performance yardstick for managed funds.

In the five years since its inception, GiveDirectly has transferred over $140 million to individuals, mostly in rural Kenya and Uganda. More recently, GiveDirectly has launched a major study of Universal Basic Income (UBI), the idea of giving poor people a regular, livable, unconditional sum of money. GiveDirectly is currently conducting a $30 million research project in rural Kenya, the world's largest study of UBI, according to researchers at the MIT Sloan School of Management, who will be studying results. Forty villages were randomly selected to provide every adult with approximately twenty-three dollars per month for twelve years (representing about half the average income in rural Kenya). Adults in another eighty villages will receive the same monthly amount, but for two years. Adults in a third cohort of eighty villages will receive the same amount as the two-year study, but with one upfront payment. And a fourth cohort of one hundred villages will serve as the control.

Research for the initiative will be headed by Princeton economist Alan Krueger, who served as chair of the Council of Economic Advisors in the Obama administration.

In addition to research services, GiveDirectly spun out a for-profit company in 2014 called Segovia that provides GiveDirectly—and any other customers—with the platform to transfer and monitor unconditional cash transfers. Segovia hopes to make cash transfers a common component of development assistance in resource-poor environments.

Why Digital Financial Services Are a Big Deal

Around two billion people on the planet are "unbanked," which is to say they have no access to financial services. Their transactions are all in cash. They have to hide any savings in their home. They are vulnerable to crime. They can't earn interest. They can't transfer money to others. They don't qualify for loans. Fortunately, new technologies are offering important opportunities, particularly through cell phones.

In the developing world, the best-known and most celebrated online financial service utilizing cell phones is M-Pesa, which launched in Kenya in 2007. M-Pesa allows users to deposit cash into their M-Pesa accounts (usually via the ubiquitous cell phone agents that sell users minutes all across Kenya), store money, and transfer money to others. They can also pay bills, purchase airtime, and in some cases, buy products.

M-Pesa was launched when Safaricom, a leading mobile operator in Kenya, saw that new cell phone users were "banking" minutes on their phones. Apparently, if someone had some money, it was safer to buy and store minutes than to hold cash. When Safaricom allowed users to share minutes with others, they saw people start to make payments to one another in this new "currency." So Safaricom decided to allow users to store and share not only minutes but also money, and M-Pesa was born.

The service spread quickly in Kenya and currently includes over twenty-five million active users—about the entire adult population of the country. A study of M-Pesa by MIT and Georgetown researchers concluded that between 2008 and 2014, M-Pesa was responsible for lifting two hundred thousand families (about 2 percent of total households) out of poverty.

M-Pesa has also been launched in Tanzania, South Africa, Afghanistan, India, and several Eastern European countries—to mixed success.

In addition, M-Pesa provides a financial platform for other services, such as M-KOPA, the Kenyan company that sells solar-powered systems for households lacking electricity. Payments for the system are made daily for a year through M-Pesa.

LOCATION SERVICES

What3words is a British firm trying to answer the age old question "Where am I?" In developing countries, complete maps, addresses, and even place names arc often missing. How do people describe where they are? By reciting two eight-digit global positioning system (GPS) coordinates? By listing landmarks? I once lived in a city in Costa Rica with the address "200 meters north and 100 meters west of the 'old tree,'" even though the "old tree" wasn't even there anymore. But locals knew what that meant.

What3words has a clever solution to location services. They have divided the planet into a grid of three-by-three meter squares and assigned each square a unique three-word identifier. (I'm currently writing, for example, from a beautiful corner of the planet located at searching.colonialist.suggested, a coastal community in Nicaragua called Las Peñitas.) The hope is that all businesses and individuals can start providing addresses with three simple words, which are then easy to find using a smartphone.

Other technologies are also offering powerful solutions regarding location services. Global satellite mapping services, popularized by Google Maps, ESRI, and others, are offering detailed traditional and satellite-view maps to be used in new ways. For example, vaccine researchers at the Gates Foundation analyze satellite images for regions not yet immunized—often because of inaccurate maps—in order to build accurate vaccination plans. When they see by satellite that a village isn't represented on maps and requires a visit, they can guide medical staff to "the village north over the hill."

New technologies are also helping define property rights. More than a billion households still live without secure, registered, documented, and tradable property rights to their homes. These "hidden" rights are hugely significant economically—likely exceeding $10 trillion in value. New registries are helping. The World Bank and others have invested in open cadastre systems. Drone technology helps with detailed mapping.

Even distributed blockchain technologies may become increasingly useful in establishing rightful ownership of property.

The world is now at our fingertips—if we have a smartphone and broadband.

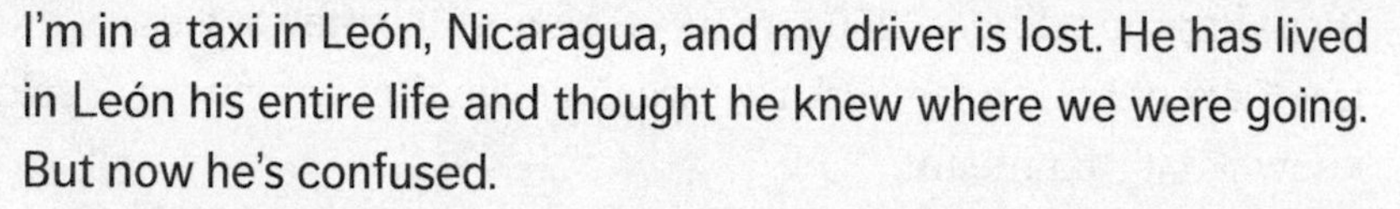

I'm in a taxi in León, Nicaragua, and my driver is lost. He has lived in León his entire life and thought he knew where we were going. But now he's confused.

So I use my smartphone to give him turn-by-turn instructions to our destination. He has never used a smartphone and doesn't know anything about GPS. He looks at me like I'm from a distant planet. In a way, I am.

When a poor country faces challenges with food, water, sanitation, health, and other life-and-death issues, it seems counterintuitive to prioritize digital programs for development. Shouldn't other priorities be higher?

Recent research, however, shows that digital solutions can both be cost-effective and enable broader progress. For example, the Copenhagen Consensus, a Danish think tank, recently consulted with the governments of Bangladesh and Haiti about the best way to prioritize development dollars. In each case, an international team of economists reviewed more than seventy possible uses of development funding, ranking interventions from most effective to least. In both cases, two digital solutions emerged in the top ten recommendations.

In the case of Bangladesh, one priority solution was the development of an "e-procurement" platform to lower the costs and corruption associated with an antiquated government procurement service. A second recommendation called for digitizing the notoriously inefficient land records system to add transparency and efficiency.

In the case of Haiti, recommendations included supporting the increase of internet coverage to 50 percent of the population—from around 4 percent—and digitizing shipment tracking systems at Haiti's largest port.

Other recommendations made by the Copenhagen Consensus were more predictable, such as investing in malnutrition programs, education efforts, and power sector reform, but the inclusion of digital solutions only highlights the importance of expanding broadband to the full planet.

From a policy perspective, let's do a thought experiment: Is it possible to have a community with extreme poverty (under $1.90 per day per person) in an environment with inexpensive, reliable broadband? On its face, this seems an absurd question: reliable broadband almost by definition tracks to communities with sufficient resources to afford it. It's a bit like saying, "Is it possible to have extreme poverty in an environment where houses also have a two-car garage?" Nonetheless, all poor communities on the planet will be getting reliable, reasonably priced broadband in the next few years. So what does that imply for extreme poverty?

Even if broadband isn't affordable immediately to individuals, it is typically affordable to government officials, health clinics, some schools, international NGOs, and others. Once broadband arrives, government programs can use identity services to reach all citizens. International efforts, including direct cash payments, are enabled. Economic advantages of better information and price data are available. Transportation services and supply chains become more efficient.

To think about it another way, for a community to really be stuck below $1.90 per person per day, it almost by definition needs to be isolated, cut off from any economic opportunities or support programs whatsoever. So is it reasonable at least to postulate that extreme poverty is in fact incompatible with broadband? That still doesn't sound right. Nobody serious will want to be associated with that idea (so we might have to coin the phrase Cashel's law).

If a household has electricity and internet, it can link to information services, education resources, health guidance, government programs, and other services. It's crazy to say that broadband access is incompatible with extreme poverty—but for many reasons, it may be true.

Even in the most remote corners of the world, it is possible to get a Coke. This has been true for decades. I remember driving past the Coca-Cola bottling facility in Blantyre, Malawi. It wouldn't have been out of place in Hamburg.

Coca-Cola is tapping its logistics and distribution expertise to bring something new to resource-poor environments: broadband. It's planning to build Wi-Fi hotspots across sub-Saharan Africa and Southeast Asia. In partnership with Intelsat, Coca-Cola is launching its "Ekocenter" program to promote local development and community. Each Ekocenter will provide local Wi-Fi as well as power and clean water.

The prototype Ekocenters look like smart, red shipping containers. A window folds up, turning the box into a small store. There are solar panels on the roof and an antenna for Wi-Fi. Tables with umbrellas and chairs create a small café. A widescreen TV plays football matches or other entertainment. The company wants to place these Ekocenters in the poorest communities in sub-Saharan Africa and Southeast Asia first, with future expansion into Latin America. When possible, Ekocenters will be run by women, which is consistent with Coca-Cola's "5by20" goal of empowering five million women by 2020.

It's all very impressive in design. I have to think that a poor, rural community getting a new café with Wi-Fi, clean water, refrigerated drinks, and widescreen TV will be an enormous hit. The Ekocenters are expensive to operate, but people I've spoken with who are knowledgeable about the effort say that Coke sees this as a consequential initiative, is committed to the investment, and views this as a way of giving back.

The ideal way to get a preview of what is about to happen with The Great Connecting would be to find a country, ideally poor and isolated, that suddenly went from very limited connectivity to very high connectivity. What would happen if that occurred?

There actually is such a country and an answer to that question.

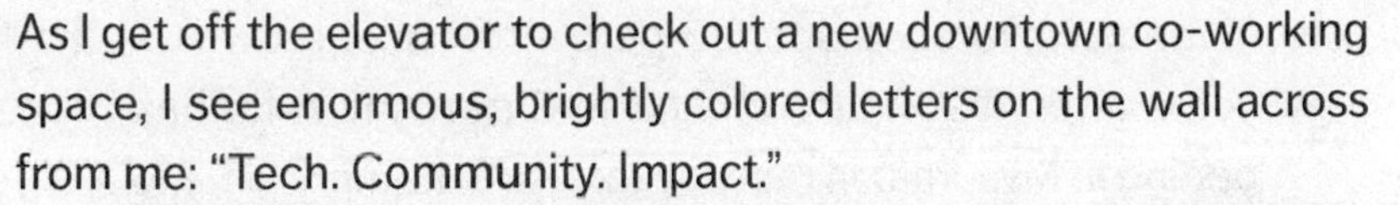

As I get off the elevator to check out a new downtown co-working space, I see enormous, brightly colored letters on the wall across from me: "Tech. Community. Impact."

Inside the space, I immediately feel right at home. It is a sprawling area with teams of techies pounding out code. There are whiteboards covered with unintelligible flowcharts. There's a maker space with a 3-D printer. There's even a Red Bull–branded refrigerator in the kitchen area. The Ping Pong table in the large common area appears to get a lot of use.

What is unusual about this co-working space is that it isn't in San Francisco or Chicago or Brooklyn. It is in Yangon, Myanmar, until recently one of the most isolated and poorest countries on the planet.

The space is called Phandeeyar (meaning "creation place"), and it is emblematic of a rapid and radical change happening in Myanmar. Just a few years ago, Myanmar was highly isolated, with the military government controlling and limiting all communications. In 2014, the country began to open to business and tourism. That year the government selected two telecommunications companies to enter the market: Ooredoo from Qatar and Telenor Group from

Norway. Both groups sprinted to install telecommunications infrastructure across the country. Over the course of about twenty-four months, the country went from under 10 percent phone and data coverage to over 80 percent, representing, according to *Fortune* magazine, the fastest growing telecom market in the world. Quality of service rose while costs plummeted. SIM cards dropped in price from $1,500 each to $1.50. Smartphones dropped from being essentially unavailable to being broadly available at low prices. Cell phones became ubiquitous, and smartphone penetration rates quickly rivaled those in developed countries.

Myanmar began this transition as one of the poorer countries on the planet, ranked by the World Bank at about 150 of all countries. In the three years before the telecommunications transition, Myanmar averaged about 5 percent annual growth. In the three years after, that figure jumped to 8 percent. A lot has been happening in Myanmar in recent years, so it is impossible to assign this boom simply to the internet. That said, studies by the World Bank, the National Bureau of Economic Research, Imperial College, and others show significant increases in GDP and employment in developing countries following the introduction of broadband.

It's easy to visualize why internet access can help push along an economy; I just need to look around Phandeeyar. To my right I see a group of about a dozen Burmese millennials huddled around a table with their laptops, building a new recruitment website. Next to them is a team working on a food delivery service. I've seen new apps for hailing cars and buying books and arranging travel. Dozens of groups around Phandeeyar are focused on their Next Big Thing. It looks like a slight majority of the professionals working here are women. I suspect the average age is younger than Taylor Swift.

The large event area at one end of the office hosts presentations on a broad range of topics. Tonight I am attending a "tech for peace" event. The hosts aren't sure yet if it will be in English or Burmese, but they promise to translate if necessary.

I'm struck that the internet itself seems fast and, as best as I can tell, unrestricted. YouTube is streaming effortlessly, and my surfing seems unimpeded, even around topics that the government finds sensitive.

Actually, there are some noteworthy differences in this co-working space compared to others I frequent in the US. The view here, overlooking the Yangon River and across the Dala District to the south, is much better than I'm accustomed to. The price is also attractive: I'm paying twenty-five dollars per week for my membership, which includes conference rooms and unlimited coffee. Co-working drop-in rates in San Francisco are closer to fifty dollars per day.

The energy and optimism I feel at Phandeeyar are also palpable across Yangon. I see a lot of new building and restoration under way. Many new businesses are springing up, some with names of mysterious provenance (one stylish coffee shop I saw was called "Eat Me Coffee"). The tourist infrastructure—getting visas, finding flights, booking rooms—has become very efficient. When I walk around, it seems like everyone in the city is nineteen years old and taking selfies.

One immediate impact of broadband in rural areas was on families. Many young men leave Myanmar to work in neighboring countries. In the airport, for example, I saw a line of perhaps 150 young laborers ready to board a flight to Malaysia. Families would lose touch with their loved ones, sometimes for years at a time, not knowing even if they were still alive. Now they can video conference at will.

In many very relevant ways, Myanmar is serving as a test case for The Great Connecting overall. A poor country gets broadband, the adjustment is fast, and a whole lot of activity is immediately set in motion, much of it very encouraging.

A Quick Visit to Myanmar

As I sit in this hip co-working space, I need to remind myself a bit about where I am.

The history of Myanmar, née Burma, over the last several centuries has been dominated by a succession of ruling dynasties, until three Anglo-Burmese wars in the nineteenth century established Burma as a British colony, of interest due to its jade, gems, and natural resources. Burma was a major battleground during World War II, much of the country was laid waste, and hundreds of thousands of soldiers and civilians were killed. Burma achieved independence in 1948, but in 1962, following a coup d'état, it became a military dictatorship.

The country is extremely diverse ethnically. The government officially recognizes 135 distinct groups, which in turn are clustered into eight "national ethnic races." The largest ethnic group, the Bamar, represents 68 percent of the population.

The combination of a military dictatorship and extreme ethnic diversity has resulted in decades of strife and violence throughout the country. Several of the conflicts, including the Kachin conflict, the Karen conflict, and clashes in Rakhine State, have been going more or less nonstop since 1948, with many people killed every year.

The military government until recently was very active in isolating Burma—and itself—from the outside world behind the "bamboo curtain." This led to a number of peculiar decisions.

In the 1980s, for example, military dictator Ne Win suddenly eliminated the national currency (wiping out the savings of millions) and replaced it with a new currency based on multiples of nine rather than ten, which he thought would bring good luck. Not surprisingly, it brought chaos.

In 1989, the government changed the name of the country from "Burma" to "Myanmar" and the capital "Rangoon" to "Yangon"—but much of the world ignored the changes due to the perceived illegitimacy of the military rulers. The name of the country therefore also became entangled in confusion. (The official government position today is that foreigners can use whichever name they want, and senior government officials make a point of alternating names to keep all factions happy.)

In November 2005, the government abruptly moved the capital from Yangon to a new city hastily built in the jungle called Naypyidaw, which led to further perplexity.

The government has also chosen to stagger the country's time zone by thirty minutes from its neighbors (MMT—Myanmar Standard Time), which confuses pretty much everything.

Isolation and international pressure on Myanmar have taken their toll. Finally, after nearly fifty years of unquestioned authority, the military government allowed general elections in 2010 and the following year nominally granted control to a new civilian government. They released Aung San Suu Kyi, the daughter of the "Father of the Nation," Aung San, from over fifteen years of house arrest; released another two hundred political prisoners; and took additional steps to improve standing with the global community. This led to the easing of international economic sanctions, additional foreign investment, increased tourism, a Nobel Peace Prize for Aung San Suu Kyi, and other benefits stemming from improved relations.

Why did the government finally relent and begin to open things up? There were undoubtedly many factors, but in one (perhaps apocryphal) story, a Burmese general was traveling in neighboring Laos and Cambodia and seeing a level of economic progress Myanmar wasn't enjoying. When he saw cab drivers with cell phones—an extreme luxury in Myanmar at that point—he realized things needed to change.

As I ride around Yangon in taxis, I'm confused. Traffic in the city drives on the right, unlike most of Myanmar's neighbors (Bangladesh, India, Laos, Thailand). Driving on the right is fine, although most of the cars appear to be from neighboring countries, so the steering wheels are (typically, but not always) on the wrong side.

Drivers and pedestrians apparently have grown comfortable not knowing which side of the car the driver is on.

Most of the buses appear to have left-hand steering wheels, as they should. Those that don't have their exits on the wrong side of the bus, meaning passengers are released not on the sidewalk but into traffic.

Myanmar reflects numerous idiosyncrasies, many due to the fact that until only recently, it was extremely isolated. Now, however, it's extremely connected. A fascinating transition is under way.

A quick search on the Apple App Store for "Myanmar" shows that mobile developers have been hard at work. I see Burmese and English apps specific to Myanmar focused on (among other things) news, travel, finance, music, entertainment, shopping, utilities, navigation, books, and reference materials.

I suspect twenty-four months ago there was nothing. What will there be twenty-four months from now? What will the techies in Phandeeyar—and the other new Phandeeyars sure to spring up—be working on next?

When traveling to parts of the world without broadband, I invariably pass through locations with an abundance of broadband. Today I'm in the massive Changi International Airport in Singapore, whose terminals boast koi ponds, a vast selection of cuisine, and soon an indoor rainforest complete with a ten-story waterfall. It also boasts broadband access: free Wi-Fi, free internet terminals everywhere, free business centers, and phones and SIM cards for purchase. How is this much abundance possible?

Later today, I'll be in Ngurah Rai International Airport in Bali, which has broadband offerings comparable to those in Singapore.

Is it possible to have too much broadband? Many visitors actually come to Bali on vacation to escape their hyperconnected lives. At the automobile exit from the airport, a hundred-foot sign with a picture of a beach and ocean reads "Let's Wander Where the Wifi Is Weak."

Broadband expansion in developing countries can be transformative. In Myanmar, the rapid introduction of broadband has brought many benefits.

Unfortunately, that is not the full story.

CHALLENGE

I'm strolling down the longest beach in the world, a seventy-five-mile strip of sand here in Cox's Bazar, Bangladesh.

"Strolling down the beach" probably evokes the wrong image, since this is the beach Bangladeshi style: rickshaws, horns blaring, women covered from head to toe (often including faces), a goat picking through a pile of trash, and a gazillion people as far as I can see in either direction. I read that Cox's Bazar mostly attracts Bangladeshi tourists. I don't think that's right. I believe it *only* attracts Bangladeshi tourists. It appears possible I'm the only foreigner here today.

The families sitting around food, the teens playing football, and the young couples sharing the beach lounges all radiate a carefree sense of tranquility. They are on vacation, and it shows. The tranquility is just a facade, though, because not many miles behind where I'm strolling, the situation is anything but serene.

In the hills behind town is a recently expanded refugee camp, now the largest in the world, full of Rohingya men, women, and children escaping persecution and violence in neighboring Myanmar. The situation in the camp is dire: over a million people live there in shelters constructed of bamboo and plastic tarps, seven hundred thousand of whom suddenly appeared a few months back, arriving with limited possessions, in fragile health, and severely traumatized from recent events.

The situation in Myanmar's Rakhine State, across the Naf River from Cox's Bazar, is even more dire. The Rohingya population there has been attacked by the Myanmar military, along with radical gangs, who have been killing the men, gang-raping the women, and burning and bulldozing their villages. The population has had no option but to escape from what most of the world describes as a textbook case of ethnic cleansing.

There are many roots to the problems between the Rohingya and the majority Bamar in Myanmar. The populations differ in religion, language, culture, and appearance and have been in active conflict since Burmese independence in 1948. There have been periodic major flare-ups in violence, leading to large numbers of Rohingya fleeing for safety at other times in the past.

This violence, however, is the worst it has ever been. There are many, many causes to the conflict. But there is one precipitating factor that everyone points to as making the situation much worse this time, one that the United Nations itself describes as playing a "determining factor" in the terrible recent violence: Facebook.

As Facebook burst onto the scene in Myanmar, it quickly became the dominant platform for both media and communications in the country. Even senior government officials communicate with constituents every day principally through Facebook.

Facebook has been used for many beneficial purposes across the country. It provides information, supports businesses, gives organizations an easy online presence, and is fun (and addictive).

But Facebook was also quickly deployed as an ideal new tool for extremists, especially radical Buddhist monks, to spread their messages of hate and violence toward minority populations. Myanmar, with its 135 ethnic groups and many decades of ongoing hostilities and warfare, had plenty of animosity ready for Facebook to amplify.

One prominent monk, Ashin Wirathu, was imprisoned for a decade by the Myanmar government for his messages of intolerance toward Muslim minorities in Rakhine State. After he was released from jail in 2012, he was introduced to Facebook and quickly realized how efficiently it could spread messages of conflict and violence. He began posting online, promoting the expulsion of the Rohingya from the country. His many messages were immediately forwarded by large numbers of followers. Graphic pictures of supposed Muslim brutality were especially effective in fanning the flames of hatred. He had many powerful helpers. The *New York Times* reports that the Myanmar military employed over seven hundred staff to further escalate hate speech online.

Facebook wasn't designed as a platform for hate speech, but it has proved, unintentionally, to be remarkably effective at reinforcing hatred online. Facebook's genius is that it can take a little bit of activity in an online community and amplify it loudly and in very targeted ways in order to promote more activity. When someone does anything on a Facebook account—posts something, comments on a picture, likes something—those actions are instantaneously broadcast far and wide. Even if someone doesn't do anything at all, activities are broadcast: for example, if somebody comments on a message thread in which I was previously mentioned, I can be alerted by text message to that comment even though I haven't done anything. Facebook has built a brilliant, finely tuned platform for amplifying human interaction.

Early in its existence, Facebook also recognized that inflammatory commentary was one type of activity that really engaged users. People are much more likely to get involved—and stay involved—if they are riled up. Those angry people will also find other angry people to associate with online. Facebook is designed in a way that aggregates like-minded people and excludes others.

The small percentage of people who are particularly aggravated about an issue represent a huge amount of page views. Facebook makes money from advertising, so having users on the site a lot, viewing a lot of pages, is good for Facebook's bottom line. It actually is *necessary* for Facebook's bottom line—Facebook doesn't as of yet have other revenue streams that compete with advertising in a meaningful way.

So Facebook is brilliant at giving a user a voice and is particularly effective if that user is upset about something. If a Little League umpire made a bad call that upset me, the community will know: I posted photos! Check out my video! Like my post! We can vent together!

Now, what happens when you introduce Facebook, with all of its masterful fine-tuning to amplify dissent, into a community riven with centuries of ethnic hatred?

It becomes weaponized.

Countries such as Myanmar have learned this the hard way. So has Facebook. In April 2018, Mark Zuckerberg said, "We're an idealistic and optimistic company. For the first decade, we really focused on all the good that connecting people brings. But it's clear now that we [Facebook] didn't do enough. We didn't focus enough on preventing abuse and thinking through how people could use these tools to do harm as well."

Why Is Facebook So Influential?

I'm in southern Bangladesh, but I'm spending a lot of time talking about Facebook. How is it possible that a technology company based in distant California holds such sway over so much of the planet?

Facebook is obviously a popular service, with over two billion monthly users and growing quickly. But how does Facebook translate a large user base into company wealth and influence?

The answer is user data. The more detailed the data, the more valuable they are to those who want to buy access—mostly advertisers.

So how is it that user data can be so consequential? Here is an example of just one form of data the Facebook now controls that can (or could) be used for profit.

Facebook is applying increasingly sophisticated facial recognition capabilities to its service. These capabilities can be really useful and fun: if I post a photo, Facebook can help identify those in the photo and even notify them that there is a new photo of them online. It's not so different than what the photo management software on your laptop probably already does.

The difference is that this service extends across all two billion Facebook users globally. Users, in turn, are posting billions of photos and tagging themselves and many of the people in the photos, which helps train Facebook's facial recognition algorithms.

So Facebook, simply because it owns "user data," starts to have the ability to recognize millions of people on the planet. (Users can "opt out" of allowing facial recognition on Facebook privacy settings, but it's a little complicated and probably few bother.)

So now that Facebook can recognize anyone, should I be allowed to post a photo of a stranger to my account and have Facebook identify the person? Maybe for a fee? Can I take a picture of a girl at a party and have Facebook identify her for me?

Hypothetically, should a police department be able to post images of suspects and have Facebook identify them? (Police departments have set up fake accounts on genealogy sites and successfully identified suspects in cases closed long ago, such as, recently, in the Golden State Killer case in California.)

Should Facebook be allowed to sell access to its global facial recognition database to third parties? Corporations? Governments? Facebook is a for-profit business. There are a lot of temptations. Recently Microsoft president and chief legal officer Brad Smith posted a lengthy blog post about the perils of facial recognition technology and called on Congress to regulate it.

The point is that Facebook has more information about more people than any organization on the planet. That confers a great deal of power.

It also confers responsibility—but given that Facebook is relatively new, growing quickly, and transnational, there isn't any one entity that it answers to. It doesn't, amazingly, even really answer to shareholders, since Mark Zuckerberg personally controls enough "Class B" shares to represent about 60 percent of voting rights overall.

So Facebook has an enormous amount of data about all of us who use the service. And it has succeeded in leveraging that asset into a prodigious amount of wealth and influence.

The Rohingya in western Myanmar are an ethnic group that doesn't exist in official Myanmar government law. Despite having lived in Rakhine State for generations (many trace their residence back centuries), the Rohingya are viewed by the government as illegal immigrants from neighboring Bangladesh. As such, they are denied citizenship, public education, state services, and civil service jobs. They are not allowed to leave their respective villages by day and are under curfew by night. The government has even prohibited using the term *Rohingya*, substituting it with *Bengali*.

Tensions between the Rohingya, who are Muslim and speak a dialect similar to Bengali, and the majority Bamar, who are Buddhist and speak Burmese, have existed for as long as people can remember. The Rohingya have faced numerous military crackdowns through the years, including in the 1970s and 1990s, both of which led to large numbers fleeing their homes. Rohingya refugees have lived in Bangladesh for decades.

Nothing in the past, however, was comparable to August 2017, when "clearance operations" by the Myanmar military forced seven hundred thousand Rohingya to escape to safety. International human rights organizations describe the operations as ethnic cleansing, involving summary executions, gang rape, torture, and total destruction of Rohingya villages. The Rohingya, already among the poorest and most persecuted populations on the planet, arrive in Bangladesh hungry and sick, with little more than the clothes on their backs. Typically they have witnessed family members shot, burned, or hacked to death by the military or radical Buddhist mobs.

The Myanmar government, and particularly the de facto head of government, Aung San Suu Kyi, has been harshly criticized for keeping mostly silent in the face of genocide. The government's inaction suggests it is fine with the expulsion of the Rohingya. Much of the Burmese population apparently is. Popular animosity toward the Rohingya, fueled by Facebook, is widespread, providing the military with both cover and support for its operations.

It is ironic that the name of Myanmar's largest city, Yangon, translates to "end of strife."

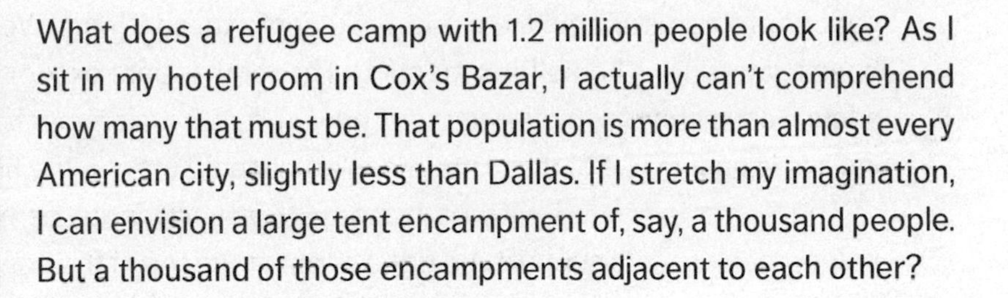

> What does a refugee camp with 1.2 million people look like? As I sit in my hotel room in Cox's Bazar, I actually can't comprehend how many that must be. That population is more than almost every American city, slightly less than Dallas. If I stretch my imagination, I can envision a large tent encampment of, say, a thousand people. But a thousand of those encampments adjacent to each other?

Let's do a thought experiment.

Imagine the community where you live. Now envision an expansive open space somewhere in the region—maybe a regional park or agricultural area or open hills in the distance.

Now imagine that seven hundred thousand poor, tired, sick, hungry, traumatized refugees suddenly show up.

How would your community respond? Or your region? Or your country?

If you were in charge of addressing the situation, of establishing response and proper policy, what would you do?

Today I'm in the middle of the Kutupalong Rohingya refugee camp outside of Cox's Bazar, the largest refugee camp in the world. Despite standing here and using my own eyes and ears and nose and empathy, I'm still struggling to comprehend this place. I see homes constructed of bamboo and plastic climbing to the ridgeline in front of me, but I can't see the twenty ridgelines of homes after that. I see hundreds of men, women, and children around me, but I can't see the million others in the area. The dimensions of the settlement defy my comprehension.

So I decide to just try to grasp the situation in the one house I'm in. I'm here touring with the Bangladeshi staff of Internews, a remarkable international organization that promotes media and communications programs around the world, including in these camps. We've been invited into a home for a meeting. The room is bare except for a few plastic chairs. I'm offered one. I have the impression that having a foreigner in one of these meetings is unusual. I may be the first.

The home is framed with bamboo, with several large central poles reaching up to support the ridge beam at the top of the house. Smaller bamboo poles support the roof. A lattice of bamboo strips reinforces the walls. The roof and walls are covered with white plastic sheets with the large blue logo of the International Organization for Migration (IOM) prominently displayed. On the floor is a sturdy mat with the green logo of the United Nations High Commissioner for Refugees (UNHCR). The rafters hold a few possessions, including some garments and a small stack of cardboard, which I assume is used for beds. The walls are empty except for a modest mirror and a pair of scissors in the corner. I note that our host today has a precisely groomed beard.

The floor is flat. If someone has money, he can purchase a bag of cement, mix it with mud, and construct a floor that stays even. True foundations are not allowed, however, so even this structure could wash down the hill in a strong rain. All structures, by Bangladeshi government rules, need to be "temporary," since refugees will be heading back to Myanmar "soon." But many residents of the camp have been here for decades. I met one young man in his mid-twenties who was born in the camp.

Because structures are required to be temporary, only bamboo and plastic sheeting are allowed as building materials. Almost all homes have dirt floors. If there is a leak in the plastic sheets during a storm, the dirt turns to mud.

There are hundreds of thousands of other structures built across these hillsides, but I can only see a few dozen in this neighborhood. The closest I come to grasping the magnitude of this place is by seeing the stacks of bamboo—bundles piled into small mountains lining the road on the way in, some the size of an apartment building. I guess if seven hundred thousand guests suddenly show up unexpectedly, you need a whole lot of bamboo.

Today's discussion is about drug abuse in the camps. Everyone is speaking Bangla. Two hundred million people in the world speak the language, but unfortunately I'm not one of them, so my mind is wandering. I'm thinking about Facebook.

Around 200,000 Rohingya fled to Bangladesh over several years in the late 1970s when the Myanmar government stripped them of their citizenship and they were declared "illegal." Another 250,000 fled in the early 1990s over several years with increased persecution by the military. All this was well under way before Facebook, which wasn't founded until 2004.

But beginning in August 2017, 700,000 Rohingya were forced over several weeks to flee government rage and local gang savagery during an unprecedented spasm of violence. The anger in the military, in the surrounding communities in Rakhine State, and within Myanmar's Bamar population overall was stoked to extreme levels, mostly by social media.

So are these people sitting here in the bamboo-and-plastic house today because of some company in Menlo Park? It's hard to say. Maybe.

In mid-June of every year, for reasons that aren't quite clear, the winds reverse over the Bay of Bengal. As the gusts heading toward South Asia increase in speed, they gain moisture over the water. When they collide with the Indian subcontinent, the air rises, cools, and drops copious amounts of rain.

The monsoon season begins with a dramatic burst, typically a fierce storm lasting several days. The rest of the season generally brings daily rains, including periodic cyclones, until September, when the storms grow weak and infrequent before finally giving way to the dry season.

The monsoon has a vital, almost mystical importance in South Asia. Flora and fauna across the region depend on the heavy rains for survival and renewal. The water table requires refreshing. Farmers depend on the downpours to grow their crops to feed the masses. The snow falling on the Tibetan Plateau from the monsoon sustains the great rivers of India and China—Indus, Ganges, Yellow, Yangtze, Mekong, Brahmaputra—upon which half the world's population depends.

Monsoon rains can be capricious: sometimes it rains too much, wreaking destruction through flooding; other times it rains too little, leaving only drought. It may simultaneously bring flooding to some regions and drought to others.

The large cities of South Asia frequently flood during daily downpours; it is common to see people walking through a foot or two of water in the streets as they go about their business.

In Bangladesh, flooding is particularly common and problematic. The country is mostly a vast river delta of the Ganges, Brahmaputra, and other rivers. Eighty percent of the country is technically a floodplain. In a typical monsoon season, 20 percent of the entire country will flood. In an especially severe year, that figure can climb to 75 percent. By percentage, Bangladesh is the most flooded country in the world. I notice that a popular model of minivan here is called the "Toyota Noah."

Cox's Bazar, which is in one of the wettest parts of the country, will receive 140 inches of rain in an average year, almost all during the monsoon season. In July, the wettest month, Cox's Bazar will average nearly three feet of rain. Accompanying the rains are frequently blasting winds, carrying sheets of water horizontally into anything in their path. In Bangladesh, even the strongest buildings built upon the most secure lands can be imperiled during the monsoon. Any building on a hillside or in a valley, and not solidly constructed, is highly vulnerable.

Development agencies working in the camps are very concerned about the monsoon, as are residents. This will be the first monsoon season since the majority of refugees have arrived. BRAC, the largest Bangladeshi NGO, has a clear understanding of the potential risks in the camps. In a blog post outlining its efforts in the camps, BRAC lists many worries: "Deaths from exposure to extreme weather could cause a public health crisis. Hills could collapse as forests have been uprooted. Roads could become inaccessible, complicating the way in which services, rations and clean water are delivered. Strong winds could blow away the majority of tarp roofs and structures, including those of health facilities and learning centres. Latrines and other water and sanitation facilities could be destroyed or damaged. Desludging sites could overflow, creating a toxic mix of water and sewage over most of the camp. Flooded latrines could contaminate drinking water and trigger a major disease outbreak."

The Rohingya already face many challenges, and the monsoon season has hardly begun.

I'm sitting in a restaurant in Yangon, enjoying a savory bowl of cashew chicken with rice.

I can see outside that the gray skies have darkened further, and the monsoon rain is peaking for the day. Torrents of falling water are caroming off of the automobiles parked nearby. Small streams along the gutters have grown into larger rivers rushing across the roadway. The sidewalks are mostly empty as people wait out the downpour in dry surroundings.

There is a sturdy roof over my head and a dry floor under my feet. The pounding sound of the rain is hypnotic, and the attenuated puffs of wind through the restaurant are quite pleasant.

I finish my meal and order a latte to wait out the storm. I'm pondering how many people in Southeast Asia aren't so lucky to have a sturdy roof during the monsoon.

Aid workers in the Rohingya camps face many urgent problems. Let me describe just one: sanitation.

If you don't immediately address sanitation issues when seven hundred thousand people suddenly appear, things get really bad, really quickly. Water-borne diseases, such as cholera, typhoid, dysentery, hepatitis, and amoebiasis, can spread immediately.

In the Rohingya camps, there hasn't been enough time to build proper latrines or sewage systems, so aid agencies acquire and deploy many thousands of portable toilets. Those toilets quickly fill, so the waste needs to be pumped frequently and then dumped somewhere. The agencies build large holding ponds for the fecal sludge.

Now mix in monsoon season: if the holding ponds fill or overflow during the monsoon, waste (and disease) is sent everywhere. So aid agencies line the embankments of the ponds with sandbags to minimize the risks of flooding or landslides. They also build roofs over the holding ponds that they hope will hold up in heavy winds and rains. If they don't hold up, then the ponds overflow. It's somebody's job to maintain and repair those roofs, sometimes in a downpour with gale winds, while precariously balanced over holding ponds of fecal sludge.

And that's just one problem.

I'm sitting at a Bangladeshi police checkpoint near the Kutupalong refugee camp. There is a long line of vehicles waiting. Some of the

vehicles are late-model SUVs driven by the various aid agencies, all with acronyms emblazoned on their sides (IOM, UNHCR, UNDP, UNICEF, WFP, MSF, CAID, etc.). Most of the rest of the vehicles are the smaller green motorized rickshaws (CNGs, named after their form of fuel, compressed natural gas) that are the principal form of transport in the area.

I passed this checkpoint on the way in. They don't seem to bother with vehicles entering the camp zone. But with cars leaving, the police are being very careful. Rohingya don't have permission to leave the camp, but given their poverty, they are particularly vulnerable to human trafficking. Through deceit and coercion, many adult Rohingya are pulled into forced labor. Many young girls are coerced into prostitution.

The police allow the SUVs to pass with barely a glance, but they are very thorough with the CNGs. They ask everyone for identification papers, and they interview anyone they find suspicious. Despite their small size, CNGs can cram six or more people in the back compartment, so there are a lot of documents to be examined.

Based on dialect, it is fairly easy for the police to identify Rohingya. Words are hard to disguise. (Writing in the *New York Times* in 2017, columnist Thomas Friedman recounts that during the Lebanese civil war, certain militia checkpoints would show travelers a picture of a tomato. Depending on how they pronounced it, they would be either allowed to continue or pulled out and shot.) The police reportedly catch and return about two thousand Rohingya each month. Certainly many thousands more get through.

Many of those are young girls in the brothels around Cox's Bazar. I haven't observed these directly, in part because the security in my hotel posts signs strongly warning against leaving the premises after 6:00 p.m., advice I have mostly heeded.

Recently, however, a young reporter working for PBS NewsHour, Tania Rashid, was in Cox's Bazar covering the sex trade. Due to her age, Bangladeshi heritage, and journalistic skill, she was able to intermix with young sex workers and interview on camera a

Rohingya girl as well as her pimp. The girl revealed that she has many clients each day, sometimes as many as five at a time. Her experiences echo the stories of women and girls who have been gang-raped by the Myanmar Army. She receives about a dollar per client from her pimp.

I feel like I've strayed awfully far from the main story of The Great Connecting. I want to talk about low-earth-orbit satellites and amazing new web apps, but here I am discussing Rohingya girls gang-raped in the brothels of southern Bangladesh, possibly in those seedy hotels I see across the street from where I stay. But unfortunately it is all connected. In Myanmar, connectivity ushered in social media, quickly followed by hate speech and violence. New broadband can have many, many repercussions.

The Rohingya of Myanmar aren't the only ethnic group to have social media directed against them. The combination of ethnic hatred and social media has proved to be a toxic mix in many places throughout the world.

In rural Indonesia in 2017, rumors spread that outside gangs were kidnapping children in order to sell their organs. The grisly photos accompanying the rumors incited rage, to the point where people in nine separate villages were lynched.

Similar rumors in rural Mexico led to the killing of two pollsters thought to be part of a kidnapping gang featured in social media posts.

In India, WhatsApp messages about child kidnappers have been blamed for a number of lynchings and mob killings of outsiders to the community.

And in Sri Lanka, animosities between Sinhalese and Tamil, which played out in a civil war that lasted twenty-five years before ending in 2009, leave the communities vulnerable to outbursts of hatred. There have been many incidents of violence, leading to the burning of businesses, houses, and places of worship. In Sri Lanka, the relatively weak media and anemic civic institutions provide only a partial counterweight to Facebook's influence. For example, the government has posted exhortations from two respected cricket stars to stop the violence, with limited success. When the government was unsatisfied by Facebook's inadequate response to the violence, it took matters into its own hands, banning Facebook, WhatsApp, and Instagram from the country. Temporary Facebook bans have occurred similarly in other countries. Eventually, however, there is community pressure to restore the service.

One senior government official in Sri Lanka, in assessing Facebook's role in promoting ethnic violence, lamented, "We don't completely blame Facebook—the germs are ours, but Facebook is the wind, you know?"

I'm traveling in southern Malawi and hearing from the locals about zombies.

Last fall, a young girl died in a hospital in Blantyre due to anemia. Doctors explained to her distraught parents that the cause of her death was probably malaria. Her parents weren't buying it. Rumors began to spread that her death was actually caused by a bloodsucker, a sort of humanoid monster that quietly invades houses in the middle of the night to steal blood. Rural Malawi has a long history of myths around bloodsuckers, as does much of Africa.

The rumors spread throughout the Mulanje District, where the girl was from, and that quickly raised suspicions about anyone not from the region. A Belgian couple planning to hike Mt. Mulanje was attacked with stones while sleeping in their VW van and almost killed before police intervened. Most NGOs ceased operations. The US Peace Corps quickly pulled out of the region.

Other attacks happened across the district, resulting in many injuries and nine violent deaths. Finally, the president of Malawi issued a statement to try to calm the issue: "If people are using witchcraft to suck people's blood, I will deal with them and I ask them to stop doing that with immediate effect."

Apparently, completely denying that bloodsuckers exist is political suicide in Malawi. (As an American, before I'm too critical about Malawian politicians' dismissal of science, I remind myself that Malawi has signed the Paris Climate Agreement.)

The Mulanje District is mostly devoid of internet or cell phones, so rumors spread only by word of mouth at this point. Communications is at the speed of walking, or perhaps bicycling.

It's likely the periodic outbreaks of rumors about bloodsuckers in the Mulanje District—and the associated violence—will continue in the future. And in the future, tens of thousands of people here will have Facebook.

The Emergence of Fake Videos

The problem of hate speech and fake information is about to get a lot worse.

For years, extremists have posted photos, often falsified, to promote their incendiary views. Videos have proved to be more immune to manipulation than photos for technical reasons—it is hard to make a fake video look real.

Unfortunately, that is changing quickly.

New, powerful tools built on AI technologies are allowing individuals to quickly, cheaply, and effectively substitute faces in videos in ways that are believable. The videos, called "deepfakes," are popular in specialty communities on Reddit and elsewhere, but they are about to become much more mainstream.

Those wishing to sow dissent or chaos now have a new, incredibly powerful tool at their disposal. What happens when an authentic-looking video appears of Donald Trump screaming for a nuclear attack? How would threatened countries respond?

Soon, it will be nearly impossible to tell a real video from a fake one. The tools to identify fake videos aren't keeping up with the technology to produce them. Experts in identifying fakes still need to magnify images frame by frame to analyze shadow patterns, for example.

Schools, civic organizations, the media, and others will need to educate people about the new technology, although this will have the side effect of bringing the veracity of all videos into question.

In the world of hate speech, a technical arms race is under way. When it comes to video, the forces of extremism will soon be gaining a powerful new weapon.

I'm on the eighteenth floor of a tower in Dhaka, Bangladesh. This afternoon I'm meeting with senior leadership at BRAC, the largest and most influential organization most people have never heard of. The office window overlooks greater Dhaka. The alleys below are crowded with pedestrians and the cycle-rickshaws I dodged to get here.

BRAC was founded in 1972 shortly after Bangladesh's Liberation War. (The organization was originally the "Bangladesh Rehabilitation Assistance Committee," later changed to "Bangladesh Rural Advancement Committee," then "Building Resources Across Communities," and finally "BRAC.") The organization employs over one hundred thousand workers in Bangladesh and ten other countries in Asia and Africa, making it the largest nongovernmental organization in the world. BRAC is one of the few international global development organizations actually based in a developing country.

BRAC's oldest and best-known programs provide microfinance support, principally to women. It has made nearly $2 billion in microloans and claims a repayment rate of over 98 percent. BRAC also manages extensive programs in education, health, disaster relief, legal affairs, and girls empowerment, as well as numerous revenue-producing enterprises, such as craft shops, publishing, dairy projects, and others. About three-quarters of BRAC's $1 billion-plus budget is provided by its own social enterprises.

Historically, BRAC hasn't been involved in humanitarian relief efforts around the world, but the appearance of one million Rohingya refugees in your home country requires attention.

Today I'm speaking with KAM Morshed, a BRAC director, about communications issues in the camps. As Wi-Fi expands quickly in the camps in order to support international aid agencies, communications issues become more complex. How do refugees get information? What sources do they trust? How can they be provided with accurate information?

Recently BRAC tried to institute a vaccination campaign against cholera, a persistent and menacing threat in the camps. Nearly all refugees refused vaccination. They believed that BRAC had plans, in cooperation with Myanmar authorities, to use vaccinations as a cover for a sterilization campaign against the Rohingya. It took BRAC months of patient efforts to prove otherwise, in part by demonstrating that Bangladeshi kids outside of the camps were also being vaccinated.

BRAC wants to understand how news travels through the camps so that they can better support their many relief efforts there. Bangladeshi government authorities probably share the same interest. The camp's population—poor, angry, poorly organized—is a prime target for radicalization by outside groups. The last thing anyone wants is for the camps to start to resemble Gaza, but how can that path be avoided?

On top of all the challenges the Rohingya face—food, shelter, health, education, security, uncertainty—we have to add terrorism to the list.

Between 2014 and 2017, the number of Facebook users in Myanmar grew from two million to over thirty million—which is roughly the adult population. Because Facebook was preloaded on phones, it became the de facto internet of the country, with no other significant sources of news or information able to compete.

Many valuable activities are supported by Facebook—benefiting businesses, schools, government agencies, and others. But from the start, the platform has also been used to spread incendiary hate speech.

Civil society organizations in Myanmar have tried for years to push Facebook to greater accountability by enforcing user guidelines. While there have been periodic communications between civil society organizations and Facebook's policy team, including meetings in Myanmar, meaningful cooperation with engineering, data, or products teams hasn't occurred.

The issue of Facebook's insufficient policing of its policies came to a head not with the atrocities taking place in Myanmar but following a 2018 Mark Zuckerberg interview with Ezra Klein of Vox. In the interview, Zuckerberg described how Facebook was responding to hate speech:

> I remember, one Saturday morning, I got a phone call and we detected that people were trying to spread sensational messages through—it was Facebook Messenger in this case—to each side of the conflict, basically telling the Muslims, "Hey, there's about to be an uprising of the Buddhists, so make sure that you are armed and go to this place." And then the same thing on the other side.
>
> So that's the kind of thing where I think it is clear that people were trying to use our tools in order to incite real harm. Now, in that case, our systems detect that that's going on. We stop those messages from going through. But this is certainly something that we're paying a lot of attention to.

The civil society organizations in Myanmar that had been battling Facebook on this issue for years weren't buying Zuckerberg's description of events. In a prompt and constructive open response posted online, they wrote, "In your interview, you refer to your detection 'systems.' We believe your system, in this case, was us—and we were far from systematic. We identified the messages and escalated them to your team via email on Saturday the 9th [of] September, Myanmar time. By then, the messages had already been circulating widely for three days." The letter was signed by six organizations: Phandeeyar, Mido, Burma Monitor, Center for Social Integrity, Equality Myanmar, and Myanmar Human Rights Educator Network. An indelicate translation of their response to Zuckerberg might be "Sorry, Mark, we call BS."

The groups recommended steps that would be helpful: more moderators who speak Burmese who can review questionable content, better mechanisms for escalation of issues with Facebook, increased cooperation between Facebook and local groups, and improved transparency around systems and metrics.

To the groups' surprise, Zuckerberg personally sent a response the next day apologizing for mischaracterizing Facebook's role and outlining a number of steps the company was taking, including hiring "dozens" of moderators. The groups were happy to have the attention of the CEO but remained wary. They had heard promising things from Facebook in the past. Would this time be different?

Hate Speech as Fake News

Hate speech is a specific (and extreme) subset of fake news and misinformation online. Fake news is a major problem in developed countries and a worse problem in developing countries.

- According to Freedom House, only 13 percent of the world's population lives in countries with a free press.
- In many developing countries, citizens have lower media literacy than those in rich countries.
- In developing countries, people often display a blind trust in information that comes from new technologies.

- Many people who get information from social networks on Facebook or WhatsApp uncritically trust what others have to say.
- In many countries, there are few credible news sources or professional journalism platforms to counterbalance rumors.
- Many entities use social media maliciously, often with the backing of the state (such as Russian government propaganda efforts in Ukraine).

Unfortunately, in both developed and developing countries, fake news is far more likely to be spread than real news. Researchers at MIT reviewed 4.5 million tweets on Twitter about 126,000 stories and found that fake news traveled six times faster than real news, was retweeted broadly about ten times as often, and reached much larger final audiences.

The next three billion people to come online in the next few years will see many benefits. They will also see a profoundly high level of fake news.

I'm in a northern district of Yangon speaking with the director of a nonprofit in Myanmar that has delivered over 160,000 meal kits to fleeing Rohingya. The organization keeps a low profile, since anything related to the Rohingya in Myanmar is sensitive.

He describes speaking with terrified families on the border forced to decide between returning home to their village, where they might be killed, or trying to cross the river into Bangladesh. The river is several kilometers wide, parts of it flow quickly, and the only transport across is a flimsy raft made of bamboo and jerry cans. There are no life preservers, and nobody can swim. It is likely many people will drown.

He has also witnessed the reality in Rohingya villages in Rakhine State. Previously he worked in some of the most challenging humanitarian crises in the world, but the Rohingya situation is the only one that has brought him to tears. It is the worst he has ever seen.

When the topic turns to Facebook, his voice fills with disgust and resignation. A foreign company has come to his country, makes money from its services, dominates the media landscape, clusters the most radical voices together in a divisive way, and contributes to ethnic cleansing—and yet, he says, it is nowhere to be seen. Facebook doesn't have an office in downtown Yangon. Facebook has no employees in the country. Sometimes Facebook doesn't even answer emails from credible organizations in Myanmar who beseech the company to remove hateful content.

How do you address a dominating foreign force you believe is doing great harm to your country when it's encamped in offices half a planet away?

Should hate speech be Facebook's problem at all? What about other forms of abusive language or misinformation? Isn't Facebook just a communications channel? Is it reasonable to hold it responsible for everything bad that people say online? Should Apple be held responsible if somebody uses an iPhone for hate speech? Should the post office be held responsible for carrying hate mail?

While Facebook is just a communications channel, it is immensely effective at what it does. That requires special treatment. A paring knife and an assault rifle are both weapons, but they are dealt with differently.

So how might hate speech on Facebook be addressed? In a perfect world, a clear Facebook user agreement about what kind of speech is and is not allowed would be sufficient. Users would read the language and moderate their own behavior accordingly. To Facebook's credit, its user agreement repeatedly notes that threatening language is not allowed. It states unequivocally,

> In an effort to prevent and disrupt real-world harm, we do not allow any organizations or individuals that are engaged in the following to have a presence on Facebook:
>
> - Terrorist activity
> - Organized hate
> - Mass or serial murder
> - Human trafficking
> - Organized violence or criminal activity
>
> We also remove content that expresses support or praise for groups, leaders, or individuals involved in these activities.

That language is unambiguous. Unfortunately, users don't read the user agreement (I suspect few employees of Facebook have read it). Even if they did, it's not clear that it would have much effect on behavior, since Facebook's day-to-day enforcement of its user agreement is limited and unpredictable.

In a perfect world, governments would define hate speech and enforce laws forbidding certain online behaviors. Unfortunately, few countries, especially lesser-developed countries with long lists of other priorities, have the will or wherewithal to deal with hate speech. Furthermore, some governments turn a blind eye or actively participate in hate speech. As an American, I understand better than ever how elected officials can use inflammatory language online to serve political ends.

In a perfect world, community norms would limit hate speech. In many environments, that is a pipe dream.

In a perfect world, the private sector would recognize an opportunity to confront hate speech, much as it has in providing sophisticated technical solutions to malware and, to a lesser extent, filtering technology for pornography. But it seems unlikely that a technology start-up will be selling solutions for filtering out hate speech online any time soon. The legal, political, and cultural issues involved are too daunting.

So in the end, Facebook has a choice. It can accept hate speech, at least at some level. Or it can prioritize filtering it out. I'm certain

Facebook doesn't aspire to be in the word police game, and "prevention" is always a tough sell, but the reality of Facebook's growth and influence demands that it take on global responsibilities it didn't foresee.

Facebook already knows what to do. The company has many years of experience in Germany, which has very strict online hate speech laws stemming from its own history with genocide. Facebook has employed a team of five hundred content moderators in Berlin in order to comply with the laws. In mid-2017, the Bundestag passed additional legislation requiring social media companies to remove obviously illegal content within twenty-four hours. More ambiguous content has to be removed within a week. The legislation has teeth: fines for noncompliance can reach €50 million. Facebook has responded by more than doubling its moderator pool, adding 700 positions in Essen. Now Facebook employs 1,200 staff to moderate Germany's Facebook users.

Facebook could do this everywhere, but as a sophisticated technology company, it surely prefers online solutions that can scale to adding lots of staff. Facebook's total employee count is about 25,000. Adding global moderators for Facebook's two billion daily users, using the same ratio as in Germany, implies a moderator workforce of around 60,000—possible, but a big step. Facebook has told the US Senate it will increase moderators and cybersecurity staff to 20,000, although it hasn't said what percentage would be focused on moderation. (As a point of reference, there are about 30,000 daily newspaper journalists in the US, which means it's possible to imagine a future where more people are paid to combat fake news than to produce real news.)

Meanwhile, Mark Zuckerberg has told informed civil society groups in Myanmar that Facebook is hiring several dozen Burmese-language moderators—a big step for sure, but it's probably only a fraction of what is needed.

Is Facebook doing enough?

I'm walking through Facebook's headquarters in Menlo Park, California, and keeping my eyes open for foxes. In 2015, Facebook opened a new, spectacular headquarters designed by Frank Gehry. The building, with the unmemorable name "MPK 20," boasts the "largest open floor plan" of any building in the world. Mark Zuckerberg maintains a standard desk in the center of the floor plan.

I'm now on the roof, strolling through a lovely nine-acre garden replete with walking paths, rolling terrain, lawns, and four hundred fully grown trees. There are Adirondack chairs scattered about, views of the greater San Francisco Bay Area, Wi-Fi, and periodic whiteboards. I'm told a family of foxes has taken up residence in the park, so I am being vigilant.

I pass a juice stand free to those seeking refreshment. I choose a premade juice blend of orange, carrot, ginger, and turmeric. It tastes like it is probably good for me.

It is hard not to be struck by the surroundings: the design, style, and thoughtfulness (and obvious wealth) are impressive. Few outsiders have the opportunity to see this lovely space, and I feel fortunate.

I'm here today meeting with two senior Facebook staff who are very involved with Myanmar. Andy O'Connell works with the product policy team, which decides the rules around what is and what is not allowed on the platform. Sara Su is a product manager, working with the team that translates rules into how the platform actually operates. Both are highly experienced. O'Connell has a graduate degree from the Harvard Kennedy School and has held responsible policy positions at the State Department and Defense Department. Su earned a PhD in computer science from MIT and has worked at Microsoft, Adobe, and Google. They strike me as knowledgeable and thoughtful.

Su recently posted a report on the Facebook corporate blog called "Update on Myanmar." She writes that Facebook has been too slow in preventing misinformation and hate speech on

Facebook and that the company has now created a dedicated team across product, engineering, and policy groups in order to be more responsive and effective. The post also outlines many of the specific steps Facebook is taking to combat hate speech generally, as well as in Myanmar.

The list, as supplemented by our conversation, is encouraging:

- hiring more Burmese-language content editors, which now number sixty and will reach one hundred in coming months
- making reporting tools on Facebook and Messenger that are more intuitive and easier to use
- improving AI systems to flag questionable content, which, in combination with content editors, are catching the majority of malicious content being removed today in Myanmar
- better coordinating with civil society groups who monitor and report hate speech
- building digital literacy programs for users and coordinating with civil society groups in training efforts
- updating content policies, including making it easier to remove misinformation that might credibly lead to violence
- banning a number of users and organizations in Myanmar, including Wirathu, Thuseitta, Parmaukkha, Ma Ba Tha, and the Buddha Dhamma Prahita Foundation (No user is allowed to support, praise, or represent these individuals and groups.)
- promoting the use of Unicode-compliant fonts, building better font conversion tools, and not allowing new users to use fonts that are not Unicode compliant
- hiring the nonprofit group Business for Social Responsibility to conduct a human rights impact assessment in Myanmar, which will be made public

In addition, Facebook has established an independent research commission called Social Science One to study issues of social and political influence. Facebook is also trying to be more transparent and self-reflective, as illustrated by Su's post on Myanmar.

The company is trying to move fast. It also has far to go. I'm reminded that the company is very young. It was founded in 2004. It is barely a teenager. And while it has been combating bad behavior online since its early days, things got particularly challenging only recently. In 2016, seven hundred thousand Rohingya had to run for their lives, in part because of Facebook. That was unprecedented and served as a wake-up call for the company. Then in 2017, it became clear that hostile, resource-rich, state powers were manipulating the platform in sophisticated ways to influence elections. Facebook initially was designed by talented engineers to allow people to share nice stories and pictures of kittens. But through its own success, it is now facing a host of challenges it never anticipated. Company staff, from Zuckerberg on down, were by their own admission caught off guard and have been playing catch-up ever since. They built arguably the most consequential communications platform in history, and because of that, they now confront issues of human rights, ethics, governance, national security, terrorism, and genocide, among others. The company at this stage in its corporate growth and influence really requires a State Department level of experience and expertise to address these complex issues. (The State Department has seventy thousand employees.) Facebook is far from that level of policy sophistication. Facebook's twenty-five thousand employees are mostly capable software engineers. Will the company ever meet the challenges it now faces? It's not certain. But Facebook is clearly trying.

On the same day Facebook published its update on Myanmar, Reuters published a scathing investigative report on Facebook in Myanmar.

The detailed analysis finds over one thousand posts, comments, and graphic images online targeting the Rohingya, some as old as six years.

Reuters researchers tried, mostly unsuccessfully, to report hateful content to Facebook. The Reuters analysis also pointed to many technical shortcomings of Facebook systems. In one glaring example of how the Burmese-to-English translation engine falls short, Reuters reports that a Burmese comment reading "Kill all the kalars [a derogatory term for Rohingya] you see in Myanmar; none of them should be left alive" is translated to English as "I shouldn't have a rainbow in Myanmar."

Facebook reports solid progress in Myanmar. Reuters reports that Facebook is not doing nearly enough, and the situation is still very dangerous. Both views are undoubtedly true.

Facebook demonstrates best intentions at fighting misinformation and hate speech online. It, however, faces four formidable obstacles in this pursuit.

First, Facebook's user base continues to grow quickly, particularly in developing countries. From a policy and security standpoint, it is hard for Facebook to keep up. Also, having a network that grows to, say, twice its original size doesn't mean twice the problems. It is more complex than that. Metcalfe's law says that the value of a network is proportional to the square of the number of connected users. A network's challenges are similarly a logarithmic function of size. Facebook's network size is growing, and the complexity of the network is growing even more quickly. It is a major challenge to design and scale systems to maintain control.

Second, Facebook started as a clever online platform designed by talented engineers. It is still a technology-driven company. From a cultural perspective, Facebook leadership would probably love to be able to address issues of misinformation, hate speech, and other emergent challenges through sophisticated technical solutions. Hiring many human rights experts (or even content editors) is a departure for the company.

Third, Facebook makes money from selling ads. Curbing misinformation and hate speech is the right thing to do, but it isn't clear that will help the bottom line—it could well hurt it. When Facebook recently announced slightly reduced projections around growth and revenue, its stock dropped by $120 billion in one day, the largest ever for an American company. Even if they were committed to doing the right thing, senior staff at Facebook must also feel financial pressures pulling in other directions. Their current circumstances are nice: recent company filings indicate the median level of pay for Facebook employees is over $240,000 per year.

Finally, how will we know if Facebook is succeeding in its fight against misinformation and hate speech? To what standard should the company be held accountable? Who should monitor and enforce accountability? Facebook is currently in a difficult situation. The press (and at times seemingly everyone else) is sniping at the company, often in a reactive and anecdotal way. Recently, for example, the *New York Times* published an article about a number of terrorist deaths in Libya linked to Facebook. At the same time, on average, more than ten people in Libya die every day from auto accidents, but that isn't front-page news because cars are a regulated industry. What reasonable standards can be agreed to about Facebook's obligations and behavior so that the company can demonstrate success?

So what should Facebook, and other social media companies, do to combat hate speech and misinformation online?

After speaking with many people in many countries about this topic, including professionals in the trenches fighting hate speech in the most hostile environments, I would summarize the "wish list" I'm hearing with respect to Facebook as follows:

- **Accept responsibility.** Facebook shouldn't be held responsible for every human ill that occurs online. That said, it does have responsibilities that need to be recognized and addressed. Its service can be used as a force for profound good. It can also be weaponized. When it comes to environments awash in racial hatred, many perceive that Facebook has delivered a truckload of hand grenades and then left communities to use them as they see fit. Facebook needs to accept the responsibility that comes with great influence, especially in developing countries that are completely unable to manage what has hit them. To Facebook's credit, the company has demonstrated an increased and genuine commitment to fighting hate speech. This issue is likely to get much more complex as Facebook further extends into developing countries, so the acceptance of responsibility will be an ongoing process.

- **Employ additional moderators.** Facebook employs 1,200 moderators in Germany, a country of eighty-three million. Facebook recently said it would hire, for the first time, "dozens" of moderators for Myanmar, a country of fifty-three million. An equivalent ratio to Germany would be closer to 800 moderators. To handle this issue adequately around the globe, Facebook will need to hire tens of thousands of moderators who speak many dozens of languages. Recently Facebook has helped fund in-country fact-checking groups in India, Colombia, Brazil, Indonesia, and the Philippines—a trend that will need to expand greatly in coming years. Facebook also has to be alert to sensitivities around *which* people they hire as moderators. For example, the perception in the Rohingya community is that most people who speak native Burmese are sympathetic to extremist views toward the Rohingya. Hiring moderators with bias (or perceived bias) adds another layer of complexity to the content-moderation issue.

- **Be more transparent.** It is difficult for organizations in Myanmar to coordinate efforts with Facebook to combat hate speech when they don't know in any detail what Facebook is doing day to day with respect to Myanmar. It is understandable that Facebook needs to be a bit quiet about its efforts, but at the same time, it can't tackle problems alone.

- **Pay better attention to technology.** Better searches, automated tools, and AI may be useful in the future in combating hate speech, but they aren't yet sufficient. One challenge unique to Myanmar, for example, is that the Burmese language is most commonly rendered in a font, Zawgyi, that is not Unicode compliant. Unicode is an international standard for translating scripts into unique numeric values, but Zawgyi, a font used by 90 percent of devices in Myanmar, is incompatible with Unicode standards. This makes technical solutions around search, translation, and AI more difficult. That said, AI is progressing rapidly, and technical problems that in the past were intractable—spam, malware, porn—have been greatly constrained by technical solutions. Facebook also faces a host of unique challenges around WhatsApp. Unlike Facebook, WhatsApp supports only private, encrypted communications. It is impossible for anyone—Facebook included—to monitor it (or flag false rumors) until news spills into other networks. To complicate matters, WhatsApp is the chosen tool for illiterate users who can't type or read on Facebook but can speak and listen to and forward messages in WhatsApp. The tools and systems used to track and moderate WhatsApp hate speech therefore need to be uniquely versatile.

- **Demonstrate increased speed and flexibility in response.** Hate speech typically appears in bursts, so it is vital to have systems that can react quickly and escalate in response. Facebook is already able to "demote" pages to various degrees, thereby minimizing the level of likes, shares, and other interactions. Facebook should even design regional "kill switches" to turn off the service in times of crisis, though these undoubtedly would be a last resort. If the company doesn't limit access in emergencies, however, governments will, as has happened in China, Sri Lanka, and elsewhere. There also should be coordination among online services and telecommunications companies: if any platform is being misused, all other platforms should help with warnings and responses.
- **Enhance anticipation.** Facebook should be able to predict with high accuracy the countries where problems with hate speech are likely to emerge. It is not that hard to figure out where ethnic or religious hatred is primed to spill aggressively onto social media. Facebook should plan ahead for that challenge and be doubly vigilant in areas likely to have problems.
- **Expand educational resources.** Facebook could get ahead of problems by educating new users about online behavior through videos or tutorials and perhaps offer meaningful rewards for completion. Facebook has run educational ad campaigns before elections in India and Mexico on how to spot fake news. These sorts of efforts could be greatly expanded. Others could play this role as well, particularly the telecoms, which have immediate access to users. A major role should also be played by Google, whose Android operating system is powering The Great Connecting and could play a much bigger role through Google.org.

- **Improve customization.** Facebook recently announced changes to its news feed that favor information from friends over third parties. In theory, this could dampen the propagation of fake news. In countries with limited media environments, however, it hasn't had the same effect, since almost all information already resides within Facebook itself.

- **Advance community.** Facebook should also ask for help. If the public sees that Facebook is genuinely committed to the issue of combating hate speech, that opens all sorts of avenues for the public, civil society organizations, international institutions, and perhaps even governments to partner with Facebook in stamping out the worst online behaviors.

It's hard to argue that it is in Facebook's immediate financial interest to stamp out hate speech. If the company itself doesn't vow to take an aggressive stance on this issue, how can it be motivated by others? Shareholders are unlikely to apply pressure. Many governments won't care. The private sector isn't likely to be a factor. This leaves the public, nonprofits, and academia as the remaining voices to argue for Facebook's increased involvement. These voices should also be quick to praise Facebook when the company takes appropriate steps in combating hate speech, as it has done in some areas.

Most of all, Facebook itself needs to decide its priorities. Past memos of senior executives at Facebook have exposed an aggressive culture of "growth at any cost." Those costs now include fake news, hate speech, and other forms of social damage, so that attitude needs to be reconsidered. Furthermore, Facebook's heritage is one of a company of talented engineers building elegant and powerful online technologies. New responsibilities—moderating, dealing with governments, responding to public opinion—are a consequence of its success. The firm needs to evolve to manage its new reality.

The good news is that Facebook does have in its mission statement—and undoubtedly in its corporate DNA—the idea of making the world a better place. Its plan for doing that, however, has to grow with the company.

To Facebook's credit, Mark Zuckerberg has personally addressed issues in Myanmar. Facebook appears to care. The company is making changes. More coordination is starting to occur, at least with respect to Myanmar. Facebook has enormous wealth, power, and clout. It can start using its talents and assets in new and beneficial ways.

The challenges, however, are profound, as Myanmar demonstrates. And we are about to have the population equivalent of fifty Myanmars coming online as part of The Great Connecting. Whether Facebook can keep up, even if it commits to a large expansion of resources, is an open question.

Because of its size and influence, Facebook gets most of the attention with respect to hate speech and other forms of misinformation online. Issues that apply to Facebook, however, apply to all social media companies. Google, Twitter, Snap, and other international firms need to share responsibility. (Even dominant Chinese apps WeChat and QQ are finding more usage outside the borders of China and are facing new, complex issues.)

But what about other groups? It is not reasonable, or fair, to expect technology firms alone to address all these issues. What about nontechnology organizations?

What can others do to combat hate speech and misinformation online?

NEWS INDUSTRY

The news industry needs to do everything possible to define and promote honest, fair, and professional journalism. Providing more background information and transparency enhances the legitimacy of the industry.

Simultaneously, the industry needs to quickly and effectively identify and debunk fake news (easier said than done). The way fake news is corrected often matters: for example, researchers have shown that using video often works better than text, repeating the fake news can unintentionally reinforce it, and having partisan voices correct the news (Republicans debunking conservative news, Democrats debunking liberal news) is most effective.

GOVERNMENT

Governments need to value and promote a free and fair press. Government efforts should be careful not to censor or constrain journalists and should include programs to support and protect quality journalism. Since journalism and government are often (by design) at odds, support for journalism needs to be codified in law.

In addition, government bodies should identify and censure organizations promoting fake news. In the case of hostile activities by foreign actors, the government should aggressively identify and sanction malicious entities working to undermine societal trust in the media.

EDUCATIONAL INSTITUTIONS

Educational institutions—from kindergarten through university—need to expose students to the challenges of hate speech and misinformation and train them to be sophisticated and informed consumers of online information. This can happen in classes specifically designed for news literacy but also in any class demanding research, analysis, and presentation of ideas.

INDIVIDUALS

Individuals need to recognize that the news environment includes misinformation, that there are objectively better and worse sources of information, and that it is everyone's responsibility to enhance their own media literacy. One important initial step is to diversify the news sources one relies on in order to be a more informed consumer of many types of news.

> I'm in a land where people online rave positively about the government. Hate speech is quickly stamped out. Users exhibit discretion about what they post online on many topics. Bad behavior is quickly addressed.
>
> What is this bewildering place?
>
> I'm in Guangzhou, China, the southern city of fifteen million residents that, along with neighboring cities Shenzhen, Foshan, Zhongshan, and Dongguan, is part of one of the largest urban areas in the world. (I'm always astounded by Chinese cities. The US has 10 cities greater than one million residents. China has 160.)
>
> I first came to Guangzhou as a backpack-toting student when China opened to individual travel in the 1980s. Guangzhou in those days had absolutely no idea what to do with foreigners. On my first night in the city, the best I could do for accommodations was the laundry-room floor of a restaurant complex.
>
> Times have changed. Arriving in Guangzhou now—in my case, via the new Terminal 2 of Baiyun International Airport—is like stepping into the twenty-second century. The structures are modern,

clean, and massive. Most systems are automated—everything from my fingerprints to my body temperature was scanned automatically. It is very clear where to go and what to do. Signage is in Chinese and English. Airport staff are polite and mostly bilingual. (All of this, incidentally, is true of much of China: later in the day, I'll be in Tianhe International Airport in Wuhan, which, unbelievably, appears to be even newer, larger, and better organized than Guangzhou.)

China has its own unique history with the expansion of the internet and online services across the country. Phone penetration in China is very high, even in the most remote parts of the country. Online services are sophisticated, and Chinese technology firms offering those services are enormous. There is no question that China has benefited greatly from the emergence of an online economy. Among other things, essentially all the smartphones linking the planet in The Great Connecting are manufactured in factories in this Guangzhou region.

At the same time, because the government is threatened by truly free speech, China has been remarkably aggressive and successful in controlling online activities. The Great Firewall of China greatly limits access to international news and social media sites. Large numbers of censors—up to fifty thousand according to Amnesty International estimates—patrol the internet. And in a new and chilling turn, the government and businesses are implementing a "social score" derived from, among other things, internet usage patterns to rank and reward good behavior.

Chinese government actions in areas where there is particularly high ethnic tension, including Xinjiang and Tibet, are astoundingly controlling. Ubiquitous surveillance, frequent police checkpoints, reeducation camps, hundreds of thousands of people in detention, and spyware required on every mobile phone—all in the name of quelling ethnic tension—are the daily reality. Chinese concerns about online hate speech are understandable. We've seen the consequences in Myanmar. But is the only alternative the extreme police state that China imposes on minority populations?

As I surf the web in Guangzhou, I can pull up my hometown newspaper, I can check the markets, and I can read some of my favorite blogs about clean tech. When I click on my Facebook app, however, nothing happens. All I get is a message reading, "Something went wrong."

That's an understatement.

Facebook was growing in popularity in China until the summer of 2009, when Chinese authorities, reacting to ethnic riots in Xinjiang Province that killed 140 people, blocked Facebook and other international social media sites. The government claimed the blockage would last just a few weeks (I was living in China at the time, eager for services to return), but the shutdown persisted.

By 2009, global social media was starting to exert its influence everywhere, including China. Secretary of State Clinton and her staff were giving speeches promoting internet freedom as a "human right," decrying efforts by governments, including the Chinese, to restrict access. Once the Chinese authorities blocked international services that summer, their resolve only strengthened over time.

At the end of 2010, Mark Zuckerberg visited China for the first time, along with his girlfriend (now wife) Priscilla Chan, who is of Chinese descent. That same year, he had vowed to learn to speak Mandarin, in part, he claimed, so he could speak with relatives of Priscilla's who don't speak English.

Efforts on the part of Zuckerberg and Facebook continued in China. In 2014, Zuckerberg again toured China, including holding a twenty-minute question-and-answer session at Tsinghua University entirely in Mandarin. Mandarin speakers mostly said his language "wasn't terrible." (I've studied Mandarin, and I was impressed by his progress!) He then issued a taped New Year's greeting in early 2015 and returned to

Tsinghua University that fall to again hold court, this time with more fluent skills. He has met with senior Chinese officials, including Xi Jinping, in China and the US. He has praised Xi Jinping's book of quotes and speeches *On the Governance of China*. At a White House dinner in 2015, Zuckerberg even asked Xi Jinping to suggest a Chinese name for his soon-to-be-born first child, which in China is considered a high privilege. (Xi declined the offer.)

Yet these efforts appear to be for naught. Facebook remains blocked. Instagram, which Facebook acquired in 2012, was blocked in China in 2014. WhatsApp, which Facebook acquired in 2014, was blocked in China in 2017.

Today, China blocks all major international social media sites. At this point, China has its own highly sophisticated social media and mobile app environment dominated by Baidu, Alibaba, Tencent, and several other behemoth firms. China's internet is almost completely in its own world, disconnected from the rest of the planet.

The Chinese Communist Party is trying to thread the needle on globalization. On the one hand, it craves international markets. All students in China now learn English from grade school, and many people are admirably fluent. (When my family lived in Chengdu in 2009, my daughter's third-grade teacher would text me each night, in English, with an update about my daughter's day—can you imagine?) Chinese are traveling internationally in droves. Currently, about 150 million Chinese tourists travel internationally each year, and that figure is growing quickly. Already they spend twice as much as international travelers from the US, which in 2017 totaled thirty-eight million (eighty-eight million when including Canada and Mexico).

On the other hand, the Chinese Communist Party is petrified of not having full control of online information and discussion. At this point, it is impossible to imagine it allowing unrestricted access to the internet. That is especially true with respect to Facebook.

Facebook and Mark Zuckerberg have worked tirelessly to woo China. Zuckerberg is clearly very invested in China. The feeling, however, isn't mutual and is unlikely to change.

Facebook has over two billion monthly users. Since Facebook's founding, the company has overcome every obstacle it has faced—financial challenges, legal threats, business models, competitors, technical issues, legislative barriers, negative public sentiment. Or more accurately, Facebook has overcome every significant obstacle to date save one: the Chinese Communist Party.

By every indication, China will continue to maintain its own walled-off internet. As the planet becomes fully wired, there is always likely to be a separation between China and everyone else.

Perhaps this book might be better titled *The Great Connecting . . . and China*.

When I speak with someone in the US about the fact that soon, three billion more people will be online, I typically get an ambivalent reaction. Americans understand the benefits of the internet, but they also see the downsides. We know all about fake news, and identity theft, and cyber fraud. We know about online addictions and anxieties. We're all constantly distracted by our devices, even as, when late for a meeting, we drive our SUVs at high speeds. (One out of four accidents in the US is now caused by texting while driving.) We seem to be more aware of the costs than the benefits of our online lifestyles.

In the US, web access, while fundamentally useful, is still an "add-on" to the rest of our lives. Even without the internet, we have access to information, education, health, banking, and other services. Although we've grown addicted to the internet, we convince ourselves we could live without it.

In resource-poor settings, however, if you have no other access to information, education, health, or banking, the internet is profoundly influential. I think we as Americans can't really appreciate that. For that reason, we greatly underestimate the significance of The Great Connecting.

I'm in Yangon, Myanmar, meeting with one of the leaders of an organization that fights hate speech online (and as such benefits from keeping a low profile). It's challenging: this guy has become a global authority on a topic he certainly wishes didn't exist.

He has a lot to say about what he would like to see happen online, particularly in cooperation with Facebook, which essentially is the internet in Myanmar.

I ask him if he has seen any progress in efforts in Myanmar to combat hate speech. He isn't sure. Hate speech tends to arrive in bursts, triggered by political events. He'll know more when the next burst arrives.

I ask him if he's optimistic about efforts in the future. He doesn't know. Myanmar is a small country far from Facebook headquarters. He understands that Facebook has higher business priorities than focusing on problems in Myanmar.

With respect to all of my questions about the expanding internet in Myanmar, he is palpably conflicted. He comes across as a bit weary.

Trying to cheer him up, I ask him about some of the benefits that the internet has brought to Myanmar. He doesn't have ready answers. He says that online banking still isn't very good. Many internet services are still very new. Not many government services are online. Local media have no chance to compete against Facebook. He confesses that he spends his days in the bad parts of the online world, so maybe he is jaded. He also says he undoubtedly has habituated to the benefits, so he doesn't see them anymore. I notice that as we speak, he is receiving a steady stream of alerts on his smartphone.

I then ask him what one thing he would be excited about for the future. I thought he might say more organizations fighting hate speech. Or better education of users in Myanmar. Or perhaps better tools from Facebook. I can see that he has something in mind that is improving his mood. He smiles for the first time.

"Unlimited 4G."

More connectivity around the globe. A fuller global union. Is The Great Connecting a good thing? Is there a path to making this partnership work?

PARTNERSHIP

I am speaking today to a Malawian development official about the coming arrival of more affordable broadband.

He is skeptical. He's heard it all before. There are too many obstacles, he says. The reality in Africa, he says, is that life in the village is unlikely to change at all. Ever.

Maybe he's right. What do I know as an outsider? Maybe I'm both naive and wildly unrealistic in believing that the next few years are ushering in profound changes to developing countries. Maybe I've spent far too little time in the poorest countries to really appreciate the obstacles. Maybe I've spent far too much time in California, immersed in the optimistic language of "network effects" and "tipping points" and "abundance." Maybe I'm biased by the fact that the new global infrastructure of communications—Facebook, WhatsApp, Instagram, Google, YouTube, Twitter—is all based where I live and work. Maybe I place too much emphasis on the future of innovation and investment. I read recently about one new venture capital fund called the Vision Fund, which has $100 billion to deploy into new digital start-ups. (Yes, I double-checked the zeros.) Maybe none of this will ever have any impact on the women of Mutambi.

But a few things are certain. I know that developing countries are becoming electrified. Prices of solar energy are dropping. The prices of batteries are dropping. Every corner of the planet will have at least some electricity relatively soon.

Cheaper broadband is also coming—that is certain. There is too much innovation and too much investment to conclude otherwise. The timeline is a little unclear—but frankly, it's possible it all might actually take less time than the experts anticipate.

Maybe I'm naive. Maybe I'm unrealistic. But I think The Great Connecting is happening over the next few years—not that many months in total, actually—and I'm willing to bet I'm right.

I see in the press that for the past three years, NGO Advisor, a Geneva-based independent media organization, has ranked BRAC the top NGO in the world based on impact, innovation, and governance. That's impressive. I also admire the full top ten list of global NGOs: (1) BRAC, (2) the Wikimedia Foundation, (3) Acumen Fund, (4) Danish Refugee Council, (5) Partners in Health, (6) Ceres, (7) CARE International, (8) Médecins Sans Frontières, (9) Cure Violence, and (10) Mercy Corps.

This list reads like a short list of future Nobel Peace Prize winners (actually number eight already won, so that leaves nine to go). I find this very inspiring.

The Rohingya refugees in Bangladesh are restricted to their camps. They don't have permission to enter anywhere else in Bangladesh. They don't have permission to access government services or even legally buy goods requiring identification, such as SIM cards.

In surveys, Bangladeshis have been asked what they think about allowing the Rohingya into the rest of Bangladesh—in essence, opening the border and providing legal status.

A majority of Bangladeshis say yes—it is the right thing to do. I find this so inspiring.

Investment and innovation are coming. Many organizations are working hard around the world to assist developing countries. Bangladeshis entertain the idea of accepting 1.1 million refugees into their country.

There is a lot to be optimistic about. Maybe sometimes progress happens more unexpectedly and more quickly than you anticipate, even in surprising places.

I'm enjoying a lovely spring evening on the outdoor terrace on the 148th floor.

I didn't know there was such a thing as an outdoor terrace on the 148th floor—but there is in one place on the planet: Dubai's Burj Khalifa, the tallest building in the world. I'm up about 1,800 feet, and the view seems infinite. Remarkably, there is another 900 feet of building and spire above my head. That's the equivalent of an additional eighty stories.

The building is not only tall; it's exquisite. The lines of the structure, all designed to stretch as close to the heavens as possible, are mesmerizing.

Dubai is a city of superlatives: the tallest building in the world stands next to the largest mall in the world, the Dubai Mall, which boasts 1,200 stores and 200 restaurants. (I noticed that Cheesecake Factory, Cinnabon, and Five Guys were all very popular.) Next to the Dubai Mall is the largest fountain in the world, shooting water five hundred feet in the air. (When the fountain opened in front of the Dubai Mall, the developer held a nationwide contest to name it. The winning choice was the "Dubai Fountain.")

Dubai is also amazingly diverse. Large murals near the building exits of those responsible for building Burj Khalifa reflect every skin color on the planet. My flight on Emirates, Dubai's flagship carrier, included a crew representing twenty-one countries and speaking twenty-two languages.

While I gaze down half a kilometer from the outdoor terrace at the astounding development below—which includes, incidentally, an indoor skiing mountain with real snow—what is particularly striking is how recent this all is. In the early nineteenth century, Dubai had a population of about seven hundred total. Dubai's first hotel didn't open until 1959, its first asphalt runway in 1965. About that time, oil and natural gas were being discovered in the region, but mostly by Dubai's neighbors. Dubai has few natural resources and makes its money from its large ports, airport, financial services, and tourism.

If anyone back in the 1970s had predicted that in less than two generations Dubai would boast the world's tallest building and the world's only seven-star hotel, that person would have been ridiculed. The Dubai airport, which transports more international passengers than any airport in the world, has thirty double-decker aircraft loading up at any one time to carry travelers to every corner of the planet.

Two generations ago, Dubai was an afterthought. Today, it feels like the future. It is a good reminder: miracles happen. Anyone visiting Singapore understands that. Or Shenzhen. Or certainly Dubai.

There is much we can do to enhance the positive and prepare for the negative as The Great Connecting approaches. We can take advantage of this key moment in human history.

I've talked a lot about Facebook and hate speech. I've talked about the responsibilities of other groups to combat misinformation online.

But what about steps we—governments, nonprofits, academic groups, individuals—should take to prepare for The Great Connecting?

RECOMMENDATIONS

1. **Anticipate The Great Connecting.** The first recommendation is simple: acknowledge that The Great Connecting is really happening and happening very quickly. That has profound implications for the entire planet.

I remember not too many years ago talking with organizations about this idea that "the internet is coming." They would debate whether they needed a website. Now, for many of those same organizations, their website effectively *is* their organization.

A few years ago, Mark Zuckerberg said that "mobile is coming" and that Facebook would focus on "mobile first." Now Americans check their phones twelve billion times a day according to Deloitte. Ninety-five percent of Facebook's global users access the service by smartphone.

Today, The Great Connecting is coming. What does that imply for your organization? If you are in global development, should you be focusing on digital programs that can scale? For those in commerce, should you focus on product offerings for billions of new customers? In global affairs, should you reach citizens directly without using government or media intermediaries? The game is changing, and we've hardly begun considering the opportunities.

While we make plans for our own organizations, we should also be making plans to welcome and orient newcomers to broadband. A "welcome wagon" program, which educates new users about the benefits and perils of online services, could help promote the good and limit the bad. A service, in the appropriate language, that highlights best resources, addresses key challenges, and offers avenues for further assistance will help speed the learning process for new users. The program would be developed and managed by either governmental or international organizations to avoid favoring any given corporation.

2. **Facilitate the adoption of broadband.** Governments and international development agencies should focus on accelerating The Great Connecting through the identification and elimination of barriers. Cost of bandwidth can be addressed through free hotspots. Lack of electricity can be addressed through subsidized programs promoting solar. Confusion about the value of the internet can be addressed by providing high-value government services online.

   Priority efforts should include identity services, online banking, land ownership registries, and other foundational government services and benefits, especially those related to health and education. E-government can be an effective carrot in promoting broadband adoption.

   A particularly effective way to make all this happen quickly is to support—even through subsidies—the tech incubators that develop in bigger cities. They play an outsize role in communications, training, and promoting new technologies.

3. **Mitigate online challenges.** We already know the many ills of online life: fraud, identity theft, porn, fake news, hate speech . . . It's a long list. As a global society, we need to prepare for the inevitable problems as three billion new users join us online.

   The highest priority should be implementing systems of credible news and information to provide quality communications and, importantly, serve as a balance to the inevitable fake news and hate speech that will appear. Independent and credible media desperately need help to take root in the new environment.

4. **Regulate as required.** Having the entire planet online increases—or, rather, exaggerates—the many online challenges we already face. Global institutions aren't currently designed to deal with topics such as privacy standards, security requirements, and the homogenization of national policies. The large internet firms have a role to play. The telecoms have a role to play. Governments

have a role to play. But realistically, we need an international "digital tribunal"—perhaps sitting in Geneva or The Hague and perhaps funded through corporate fees—to serve as a shared resource for policy and dispute resolution. This is uncharted territory.

5. **Celebrate The Great Connecting.** Can we choose a milestone to signify completion (or near completion) of The Great Connecting?

   When the first transcontinental railroad was completed in the United States in 1869, dignitaries gathered at Promontory Summit, Utah, and drove a golden railroad spike into the meeting of the two lines. It was an event of global consequence.

   When the Suez Canal was completed, also in 1869, world leaders gathered for a celebration involving cannons, fireworks, and flotillas of ships. It too was an event of global consequence.

   What would an appropriate celebration of The Great Connecting look like? Global parties? Government proclamations? Maybe we need to take a global, planetwide selfie on some chosen date in the next few years?

The Great Connecting is happening. We can make it happen sooner. We can make it better. We can celebrate the first true global union for the planet.

So, to summarize, the following are our priorities in preparing for The Great Connecting:

- anticipate
- facilitate
- mitigate
- regulate
- celebrate

What would be the three most important specific next steps? Here are a few reasonable candidates:

- better resources for newcomers
- more coordinated cooperation with (and pressure on) Facebook from governments, nonprofits, and academia to address hate speech and other forms of misinformation
- establishment of a global governance council to address global connectivity issues

There is a lot to prepare for. Time is limited. It's possible that The Great Connecting will happen even more quickly than anyone anticipates.

A few years ago, a Yale college student was home visiting her father during a school break. She had assignments due and complained to her father, "Dad, the internet in our house sucks."

The college student's father, Mukesh Ambani, is chairman of Reliance Industries, a petrochemical consortium in India and the country's richest person. As he tells it, his daughter's complaints, as well as his country's poor internet infrastructure overall, prompted him to start the largest, fastest internet expansion in India's history.

Reliance Industries launched Jio, a telecommunications firm, and promptly spent $35 billion building a high-speed 4G cellular network across India. Jio built two hundred thousand new cell towers and laid 150,000 miles of fiber optic cable.

In 2016, Jio launched the network to consumers, offering free phone calls, free texting, and six months of free data, after which data charges were about one-fourth of the industry average. Usage skyrocketed in terms of both subscribers (now over two hundred million) and data usage (now the highest in the world for any telecommunications company).

In 2017, Jio introduced the JioPhone, a hybrid feature phone / smartphone that takes advantage of 4G data speeds. Among other features, the phone comes preloaded with five hundred streaming TV channels and music in seventeen languages. The phone is essentially free: it requires a twenty-three-dollar deposit, which is reimbursed with the return of the phone.

Josh Woodward of Google, who has led teams building new web services in India, says that thanks to Jio and the JioPhone, "hundreds of millions of users are now going to come online faster than all the models projected."

It's a remarkable story. I live in California, where countless talented Indian engineers work in technology and contribute every day to the global economy. Many major corporations are led by Indians. Google is led by an Indian. Microsoft is led by an Indian. Then I think that eight hundred million Indians don't have the internet, but they will soon. What sort of wave of talent does that herald?

We'll know soon. A few months ago, Mukesh Ambani claimed his network was still only at 20 percent capacity and that "we are determined to connect everyone and everything, everywhere."

Things are progressing quickly in India as millions of people are introduced to broadband services every week. Their gateway to the internet, however, is the smartphone, which requires users to know how to read, write, and type. Are there technologies on the horizon that might allow easier access to the internet for the next few billion users?

I'm in downtown Seattle admiring the cranes—the mechanical type, not the avian type. There are cranes hovering over building sites in every direction. The frenetic activity is mostly due to one company: Amazon. It occupies or is constructing over thirty office buildings in

Seattle to house its expanding local workforce, currently numbering over forty thousand employees. Amazon dominates the downtown, occupying as much office space in Seattle as the next forty largest employers combined. The millennials I see filing to work seem to have it good. I see a car park, a bike park, and a dog park. Many employees bring their dogs to work: I see at least seventy-five canines heading with their owners to the offices upstairs.

With all the cranes, I'm reminded of the furious building craze in San Francisco in the 1990s. I try not to take it as a bad omen that Amazon has built three enormous buildings shaped like glass bubbles, the largest ninety-five feet tall, in the heart of the construction.

I'm currently inside one of those bubbles, enjoying a latte, sitting among ferns and trees, and speaking with Emily Roberts, who is a senior manager for Alexa at Amazon. Among other responsibilities, Roberts and her team think about new uses of voice-activated "virtual assistants" in emerging markets.

There is a lot to think about because the technology is so new and is evolving so quickly. In some ways, using voice as a user interface (UI) is very constraining compared to typing and reading. However, Alexa does have a number of advantages over traditional internet devices:

- The UI is easy to understand—it is simply voice.
- Building Alexa speakers is cheaper and easier than building smartphones.
- Alexa is hands-free.
- Alexa is easier for children to use.
- Alexa is easier for illiterate and innumerate people to use.

When smartphones first appeared, developers (such as those in my company, Forum One) struggled with how to, in essence, replicate a full website on a tiny screen. But over time, they created new UI tools that worked well on smartphones. Developers also took advantage of the unique capabilities smartphones offered—portability, GPS, cameras, authentication, motion detectors—to enable phone-native apps that would offer functionality surpassing traditional computers. Before long, as we discovered, we were all hailing cars, sending selfies, and swiping right. Now the phone is every bit as powerful as the computer, just different.

The same evolution will inevitably happen with voice assistants like Alexa. The current most popular uses are relatively mundane, such as games and sports. In developing countries such as India, Roberts reports that "devotional skills," those relating to religion and spirituality, are popular.

There is certainly much more to come, however. In order to explore opportunities, Amazon conducted an Alexa "Tech for Good" developer contest to create Alexa skills with a positive impact on the environment, local communities, and the world. The winning entry, "World Mathematics League," allows users to participate in fifteen minutes of daily math questions.

Amazon also offers an Alexa Skills Kit (ASK), providing tools, documentation, code samples, and other resources to assist developers in building new skills for Alexa. In addition, Amazon offers an Amazon Voice Service (AVS) to allow hardware manufacturers to embed Alexa in their products (such as cars or TVs).

In developing countries, there are many hypothetical uses for Alexa. Even having a single-use service—"Alexa, tell me the weather tomorrow"—would be revolutionary. Could Alexa speakers help with schooling? Reading to children? Accessing government services? Diagnosing diseases? The fact that illiteracy rates in much of the developing world are still over 50 percent means that hundreds of millions of people can't access the internet through current devices. They can with voice technologies such as Alexa.

Speakers are also getting cheaper. It's possible to imagine in many environments that someday soon speakers will be handed out for free. That has already happened in India with phones, which are generally much more complex to manufacture than speakers.

We are very early in the game, and nobody has yet figured out the role of smart speakers in resource-poor environments. Many, however, are trying. There is a high-stakes race under way, and Amazon, as a major participant, is investing heavily. Jeff Bezos has said that Alexa may become the "fourth pillar" of Amazon, joining retail, Prime, and AWS.

When you combine Alexa with Starlink with a solar panel with a poor hut in Malawi, what do you get? We don't know. We will soon.

The Race for Virtual Assistants

Despite the massive power of the internet, the way we interact with it is very limited. We type—which is slow and cumbersome—and read, which is difficult in many contexts (and assumes you can read).

Great progress has been made in the last few years on the next generation of internet user interface: voice. Soon, much of our interaction with the internet will be through conversation.

There are four major entrants in the race to become the world's dominant virtual assistant: Amazon Alexa, Microsoft Cortana, Apple Siri, and Google Assistant. Industry experts believe that whoever controls "conversation" online and serves as the gateway to online services will command a very powerful (and lucrative) position. The virtual assistants are being integrated with our phones, computers, and countless other devices. They are also incorporated in free-standing "smart speakers" sold by the firms.

Researchers and journalists are fond of putting the four main virtual assistants to the test by asking each a list of hundreds of questions and then judging the accuracy and utility of the responses. The competition quickly evolves with the technology, but currently, the consensus is that Google Assistant is the smartest. Siri in the past has been described as the least intelligent of the four, but she makes up for that by being the funniest, delivering comic quips at twice the rate of the others. She is coming on strong, however: most recent comprehensive tests have Siri in second place, followed by Alexa and Cortana.

Who will win the race? It's difficult to bet against Google. It has the best expertise in search and in AI. It has a massive technical infrastructure, a large developer community, and good connections at the enterprise level. But the other three tech behemoths are investing the resources to make a competitive race of it.

Internet access in India is growing by five million consumers per week. Most of those millions are buying their first smartphones, but some are buying smart speakers with Alexa, Google Assistant, and Siri.

Amazon is making an especially aggressive push in India, in part by making Alexa adept in Hinglish (the Hindi + English dialect popular across much of the country). In addition to mastering accent and vocabulary, Alexa has learned about Indian holidays, biryani recipes, funny cricket jokes and Bollywood plot lines. Developers have added ten thousand "skills" to Alexa's repertoire that are appropriate for the Indian market. By comparison, when Alexa launched in the US, she had just thirteen skills. Amazon is working hard to increase the skills "ecosystem," paying the most successful skills developers over $100,000 annually.

By the way, Alexa in India doesn't only speak Hinglish. She also speaks other local dialects, including Hindi, Kannada, Tamil, Telugu, and Malayalam. Alexa must be doing something right: after launching in India, she received 450,000 proposals for marriage.

I'm in Dhaka, Bangladesh, watching Indian TV. Ad after ad is for Amazon Alexa. My, she looks clever—witty, insightful, and charming.

I decide to speak with her myself. I set my Alexa to Indian English.

> Me: Alexa, tell me a joke.
>
> Alexa (in Indian accent): What do they use to light up the cake at a Bollywood party? Scandles.
>
> Me: Tell me another joke.
>
> Alexa: How do you cross a one-way street in India? By looking both sides.
>
> Me: Tell me another joke.
>
> Alexa: Do you know how many food jokes I know? Naan.

The expansion of smart speakers in developing countries raises an interesting prospect. While the next billion internet users will use smartphones, the next billion or two after that are likely to use smart speakers, due to lower costs and simpler user interfaces. What does that look like? How will Alexa be perceived in developing countries?

I know the first time I spoke with Alexa (I think my question was about the weather), I was impressed—but not overwhelmed. I had been around technologies leading to Alexa for many years. As such, my thoughts were mostly contextual: "Wow, the voice recognition has gotten quite good"; "Hey, my bandwidth is holding up well"; "Nice, AWS cloud computing is really fast"; "Hmm, let me find edge cases that Alexa can't handle . . ."

I now think about the first time villagers in a hut in Africa ask Alexa a question. The villagers don't know about the internet, much less about voice recognition, bandwidth issues, or cloud computing. They don't even have much experience with electricity—or any other technology whatsoever. So how do they contextualize Alexa when they ask a question and get an informed response? Is that a person? How does she know everything? Does she really know all of the past and everything happening around us now? How does she know what the weather here will be? Can she predict the future? Can she really speak dozens of languages? Where is she? What does she look like? Why doesn't she come here in person?

I think the answer is that they aren't able to contextualize any of it. It is simply far too great a leap. I had decades of preparation for Alexa. They have minutes.

So actually there is only one way they can understand Alexa: Alexa is God.

I'm in Johannesburg, South Africa, and a young boy passes me with a shirt that reads "There is magic in the air—it is called 'WIFI.'"

I'm in a room in Mountain View, California, and in walks Elon Musk.

He's wearing a T-shirt with some sort of abstract print on the front and a badass black jacket that wouldn't be out of place on the back of a Harley. It's summer, so he must be hot.

He is here today giving a presentation and answering questions about Tesla. To be honest, he looks and sounds pretty beat. I know he is prone to all-nighters.

Unfortunately, he isn't here to talk about SpaceX. I have so many questions about Starlink and satellites and connecting three billion more people on the planet—but that's all off topic today. I want to ask him about his ambitions for global internet communications, and whether he thinks it will help alleviate poverty, and if he is concerned about the Rohingya, and if he has ever seen a hut in rural Malawi. I would love to talk with him about The Great Connecting. But he is here to talk about battery manufacturing and solar roofs and electric vehicles—not satellites and broadband.

After his brief presentation, he asks if those in the room have any questions. I'm not interested in Tesla. I'm interested in Starlink.

I ask, "Elon, Gwynne Shotwell, the president of SpaceX, has said that Tesla cars will use Starlink satellites. Can you elaborate on Starlink and when that will happen?"

He basically says they aren't sure yet and is quick to change the subject.

OK—that's probably about as much information as anyone will get about Starlink at this point. I tried.

After the Tesla presentation, I sit outside. It's a warm and clear California afternoon. I'm at the Computer History Museum in Mountain View, a sort of shrine to technology in the heart of Silicon Valley. On the walls inside hang photos of museum fellows who represent a who's who of technology's high priests and priestesses (mostly priests): Tim Berners-Lee, Gordon Moore, Steve Wozniak, Admiral Grace Hopper, and a few dozen other images gaze out at us mere mortals. If one added a few beards and hats, the framed visages might start to resemble the Russian icons I've seen hanging in the Byzantine chapels of the Kievan Rus.

In the parking lot to my right, about a third of the cars are Teslas. On the boulevard in front of me, a steady stream of imposing, unmarked buses with black-tinted windows cruise by, heading to Google headquarters around the corner. Facebook is up the road to the left. Apple's new spaceship campus is not far behind me.

If I had to choose one place as ground zero for technology and technology's promise to reshape the world, it's probably here, in the heart of Silicon Valley. And if I had to choose the precise other end of the road from the women of Mutambi, this is it—a small parking lot in Mountain View off of Pear Avenue.

The worlds of Mountain View and Mutambi aren't connected at all. But they will be very soon. The Great Connecting is happening.

I'm now home. I'm on a hill overlooking my town of Sonoma. The morning is quiet. All I hear are the tolling of bells from a church in the distance. I see below me a cluster of historic buildings at the town center. To my right, green corrugated vineyards reach to the far side of the valley, giving way to brown rolling hills. Far to the west, a bank of fog is receding toward the Pacific. To the south is the miniature skyline of San Francisco.

For some reason, as I gaze over the pastoral scene, I'm not thinking about Sonoma or the people below, waking to their daily chores. Instead, I'm thinking about relationships. If The Great Connecting represents a new global relationship, is it going to work out?

I'm not sure. Relationships require the ability to explore what's possible. I think about how broadband is reaching every corner of the planet and how people are using it in innovative ways. Relationships require investment. I think about the amazing efforts of the Hamels Foundation, the Carr Foundation, the Gates Foundation. Relationships require the ability to learn from mistakes. I think about all that's starting to happen to combat fake news and hate speech online. Relationships require an optimistic sense of the future. I think about excited millennials I met in Yangon who are building the next great online service. On balance, I think this relationship will work out. The benefits far outweigh the challenges. I'm highly optimistic. I'm also realistic: this relationship is complicated. It needs our attention.

As I stand above Sonoma, my mind wanders to a family I visited in rural Nicaragua. The grandmother was sitting in a worn and wobbly chair near the front door. (I'm actually not sure if there was a door in the front door—I saw chickens coming and going as they pleased.) I had been told that she had fallen on hard times. As the head of the household, she was responsible for three children and an infant granddaughter. Recently her foot was crushed in an accident involving her horse cart. She suffers from diabetes. The nearest health clinic is far away. She has difficulty moving around. I could see that they didn't have running water, and they couldn't afford much to eat beyond rice and beans. A family friend mentioned that the condition of the outhouse made it "dangerous" to use.

But there she was holding her granddaughter in her lap, enjoying the baby's giggles and placing a long and loving smooch on the girl's forehead. Perhaps I'm projecting as a parent, but I expect her dreams for her granddaughter are the same as my dreams for my daughters—secure health, enlightened education, a long life, lots of exploration, and countless opportunities.

And I'm also hopeful that her granddaughter will achieve all of this, thanks in part to The Great Connecting. It is the most consequential opportunity we have as a planet. We need to recognize that fact. We need to embrace the possibilities. And we need to celebrate.

AFTERWORD

It is early 2019, and as I complete this manuscript, there are headlines every day related to The Great Connecting. To name a few recent events:

- Loon has started operations in Kenya and should be providing connectivity services later this year. The company has named a prominent advisory board to help with scaling operations.
- Iridium has now deployed its fully upgraded network of sixty-six LEO satellites.
- China has launched the first Hongyun-1 test internet satellite.
- OneWeb has successfully launched its first 6 satellites of a planned 1,980 total and aims for worldwide broadband access by 2021.
- SpaceX reportedly has raised $1 billion, much of which will be used for Starlink. The company also has applied for FCC permission to deploy up to one million base stations in the US.
- Jio is still charging forward in India. Sixteen million Indians per month are gaining broadband internet access.
- Facebook is still playing catch-up on many issues. Mark Zuckerberg's 2019 personal challenge is to "host a series

of public discussions about the future of technology in society—the opportunities, the challenges, the hopes, and the anxieties."

- SpaceX's ship *Mr. Steven* is still trying to catch fairings. It is coming close.

There is clearly progress every day in wiring the planet. That is important but itself isn't the reason I left my job to research this book. I began my exploration of The Great Connecting with two key questions.

The first involved the implications, both positive and negative, of The Great Connecting in developing countries and for the planet overall. I complete this manuscript with an appreciation for a very long list of benefits and a very long list of challenges. Mostly, I'm optimistic. While future headlines undoubtedly will be dominated by problems, there will be millions of hidden, untold successes. Stories of a mom saving money, a daughter learning math, or a community being granted a voice for the first time will never make the headlines, but they represent truly consequential progress.

Second, I sought to understand what the major players involved in connecting the planet are doing to prepare for this major world event. How will they accentuate the positive effects and mitigate the negative effects of expanded connectivity? To be honest, I'm still looking for answers. If there are groups thinking hard about this issue, I'm eager to learn more. I've spoken with senior people in industry, in government, and in nonprofits, and none can point me to significant efforts. When I spoke with Tom Wheeler, who led the FCC in the Obama administration, about the prospect of billions of smartphones and devices reaching people for the first time, he communicated his concern that we aren't ready and need to start preparing: "We don't want a bunch of academics seven years from now writing books titled *What Were We Thinking?*"

The Great Connecting is happening. The good news is that we can get this right. There is still time.

APPENDIX: HOW TO GET INVOLVED

We all can be observers of The Great Connecting. It is more rewarding, however, to participate, even a little bit. How can one contribute to this epic process of global connection?

Here are seven ideas that come to mind. If you have additional ideas, you can post a comment on the blog post with the title "How to Get Involved" at the Broadband Everywhere blog (http://www.broadbandeverywhere.org).

- **Explore.** Social media brings the world to our smartphones. It is possible to meet people, join conversations, and share photos, music, or ideas with individuals around the world in nearly every country. I was once, for example, looking into a project with my daughter involving Iran. It is astounding how many Iranians use Instagram.

- **Travel.** Taking a trip to a developing country greatly supports The Great Connecting. You will use online services to research, plan, schedule, and pay for your trip. You will use your phone to coordinate plans once you are in the country. You will spend money, rewarding those who are building the information infrastructure.

- **Hire.** The Great Connecting makes it possible to hire people directly in remote places. If you or your company needs to hire a designer, developer, data entry specialist, or other worker, you can use online services such as UpWork to find people. Language tutors are especially helpful. My family has had great luck hiring instructors in Mandarin, Spanish, and Portuguese for online lessons.
- **Volunteer.** Many services allow you to volunteer online. You can tutor English, help with sister schools or sister cities programs, or be an international mentor.
- **Study.** Online education programs such as EdX and Udacity offer classes that are international. Any breakout groups you join are certain to have participants from all over the world.
- **Donate.** A number of large, credible organizations allow donations or loans to people in developing countries. GiveDirectly provides unconditional cash transfers to poor people in a number of countries in Africa. GlobalGiving supports grassroots charitable projects worldwide. Kiva provides microloans in eighty countries.
- **Participate.** New information about The Great Connecting is posted on the Broadband Everywhere blog (http://www.broadbandeverywhere.org). You can post and respond to comments there. Information will also be posted there about the best new information resources on blogs, Facebook, Twitter, and LinkedIn. We can use the hashtag #BroadbandEverywhere.

Also feel free to write me with ideas and comments: cashel @ForumOne.com.

RESOURCES

The following resources were useful in the research and writing of The Great Connecting. Some are primary sources. Others are articles discussing primary sources or newsworthy events. If a Wikipedia article is particularly informative, unique, or includes a helpful list of other articles or resources, I include it. Articles are listed in approximate order of how the information appears in each chapter.

INTRODUCTION

Ralph, Eric. "SpaceX's Starlink High-Speed Internet Satellites Alive and Well in Orbit." Teslarati.com, May 29, 2018. www.teslarati.com/spacex-first-starlink-internet-satellites-go-live-in-orbit/.

Pixalytics Ltd. "How Many Satellites Are Orbiting the Earth in 2018?" March 28, 2018. https://www.pixalytics.com/sats-orbiting-the-earth-2018/.

Russell, Kendall. "SpaceX Testifies: First Prototype Satellite Coming This Year." Via Satellite, November 2, 2017. www.satellitetoday.com//telecom/2017/11/02/spacex-testifies-first-prototype-satellite-coming-year/.

Robert Wood Johnson Foundation. "The Five Deadliest Outbreaks and Pandemics in History." March 23, 2018. www.rwjf.org/en/blog/2013/12/the_five_deadliesto.html.

Wikipedia. s.v. "List of Epidemics." Accessed June 15, 2018. en.wikipedia.org/wiki/List_of_epidemics.

Wikipedia. s.v. "World War II Casualties." Accessed July 4, 2018. en.wikipedia.org/wiki/World_War_II_casualties.

Wikipedia. s.v. "List of Religious Populations." Accessed July 8, 2018. en.wikipedia.org/wiki/List_of_religious_populations.

McGrath, Rita Gunther. "The Pace of Technology Adoption Is Speeding Up." *Harvard Business Review*, August 7, 2014. hbr.org/2013/11/the-pace-of-technology-adoption-is-speeding-up.

Anthony, Sebastian. "Smartphones Set to Become the Fastest Spreading Technology in Human History." ExtremeTech, May 9, 2012. www.extremetech.com/computing/129058-smartphones-set-to-become-the-fastest-spreading-technology-in-human-history.

ATTRACTION

Global AIDS Interfaith Alliance (GAIA). "GAIA's Impact." www.thegaia.org/.

GAIA. "Mobile Health Clinics." www.thegaia.org/gaia-programs/mobile-health-clinics/.

Wikipedia. s.v. "List of Countries by GDP (PPP) per Capita." Accessed July 8, 2018. en.wikipedia.org/wiki/List_of_countries_by_GDP_(PPP)_per_capita.

World Health Organization. "HIV/AIDS." March 2, 2018. www.who.int/gho/hiv/en/.

SpaceX. "SpaceX." www.spacex.com/.

SpaceX. "Launch Manifest." www.spacex.com/missions.

Spaceflight Now. "Launch Schedule." spaceflightnow.com/launch-schedule/.

Bierend, Doug. "SpaceX Was Born Because Elon Musk Wanted to Grow Plants on Mars." Motherboard, July 17, 2014. motherboard.vice.com/en_us/article/jp5g8k/spacex-is-because-elon-musk-wanted-to-grow-plants-on-mars.

Union of Concerned Scientists. "UCS Satellite Database." www.ucsusa.org/nuclear-weapons/space-weapons/satellite-database#.W0KXxNhKiL4.

Wikipedia. s.v. "Starlink (Satellite Constellation)." Accessed July 3, 2018. en.wikipedia.org/wiki/Starlink_(satellite_constellation).

Emre, Kelly. "SpaceX's Shotwell: Starlink Internet Will Cost about $10 Billion and 'Change the World.'" *Florida Today*, April 27, 2018. www.floridatoday.com/story/tech/science/space/2018/04/26/spacex-shotwell-starlink-internet-constellation-cost-10-billion-and-change-world/554028002/.

Forrester, Chris. "Musk's Team Sticks with Starlink." *Digital Media Delivery*, October 10, 2018. advanced-television.com/2018/10/01/musks-team-sticks-with-starlink/.

Ralph, Eric. "SpaceX's Starlink High-Speed Internet Satellites Alive and Well in Orbit." Teslarati.com, May 29, 2018. www.teslarati.com/spacex-first-starlink-internet-satellites-go-live-in-orbit/.

World Bank. "Poverty Overview." www.worldbank.org/en/topic/poverty/overview.

World Bank. "Ending Extreme Poverty." www.worldbank.org/en/news/feature/2016/06/08/ending-extreme-poverty.

Wikipedia. s.v. "Extreme Poverty." Accessed July 3, 2018. en.wikipedia.org/wiki/Extreme_poverty.

United Nations. "Sustainable Development Goals." www.un.org/sustainable development/sustainable-development-goals/.

Oxfam International. "An Economy for the 99%." www.oxfam.org/en/research/economy-99.

Credit Suisse. "Global Wealth Databook 2016." https://www.credit-suisse.com/media/assets/corporate/docs/about-us/research/publications/global-wealth-databook-2016.pdf.

Wikipedia. s.v. "Soweto." Accessed July 7, 2018. en.wikipedia.org/wiki/Soweto.

EXPLORATION

Fendelman, Adam. "An Introduction to 1G, 2G, 3G, 4G & 5G Wireless." Lifewire. www.lifewire.com/1g-vs-2g-vs-2-5g-vs-3g-vs-4g-578681.

Broadband Commission Special Session at the World Economic Forum. "Broadband Commission for Sustainable Development." www.broadband commission.org/Pages/default.aspx.

Broadband Commission for Sustainable Development. "The State of Broadband 2018: Broadband Catalyzing Sustainable Development." www.broadbandcommission.org/publications/Pages/SOB-2018.aspx.

Broadband Commission for Sustainable Development. "2025 Targets: Connecting the Other Half." http://www.broadbandcommission.org/Documents/publications/wef2018.pdf.

International Telecommunications Union. "Connecting the Unconnected: Working Together to Achieve Connect 2020 Agenda Targets." http://broadbandcommission.org/Documents/ITU_discussion-paper_Davos2017.pdf.

Wikipedia. s.v. "5G." Accessed July 5, 2018. en.wikipedia.org/wiki/5G.

Morgan, Jacob. "A Simple Explanation of 'The Internet of Things.'" *Forbes*, April 20, 2017. www.forbes.com/sites/jacobmorgan/2014/05/13/simple-explanation-internet-things-that-anyone-can-understand/#163c06931d09.

Seeds of Learning. "Home." www.seedsoflearning.org/.

Internet Society. "Brief History of the Internet." www.internetsociety.org/internet/history-internet/brief-history-internet/.

Centre for Computing History. "The Centre for Computing History." www.computinghistory.org.uk/det/2489/Tim-Berners-Lee/.

Berners-Lee, Tim. "World Wide Web." info.cern.ch/hypertext/WWW/TheProject.html.

Wikipedia. s.v. "Mosaic (Web Browser)." Accessed July 7, 2018. en.wikipedia.org/wiki/Mosaic_(web_browser).

Claro. "Claro: Teléfono celular y fijo, internet, televisión HD." www.claro.com.ni/personas/.

Huawei. "Huawei United States—Building a Fully Connected Intelligent World." April 3, 2018. www.huawei.com/us/.

Mexico Conectado. "Free Internet Access in Public Spaces." www.mexicoconectado.gob.mx/?page_id=12408.

Ministerio de Tecnologías de la Información y Comunicaciones. "El plan vive digital." www.mintic.gov.co/portal/vivedigital/612/w3-propertyvalue-6106.html.

NBN Co Ltd. "About NBN Co." www.nbnco.com.au/corporate-information/about-nbn-co.html#.

Calero, Mabel. "72 millones de córdobas se gastaron por wifi en parques de Nicaragua." *La Prensa*, February 26, 2018. www.laprensa.com.ni/2018/02/26/nacionales/2382801-72-millones-de-cordobas-se-gastaron-por-wifi-en-parques-de-nicaragua.

Broadband Commission for Sustainable Development. "Commissioners." www.broadbandcommission.org/commissioners/Pages/default.aspx.

Culturing Media. "Digital Planet." sites.tufts.edu/digitalplanet/dei17/.

Huawei. "GCI 2018: Tap into New Growth with Intelligent Connectivity." www.huawei.com/minisite/gci/en/.

Wikipedia. s.v. "List of Countries by Number of Mobile Phones in Use." Accessed July 8, 2018. en.wikipedia.org/wiki/List_of_countries_by_number_of_mobile_phones_in_use.

Transsion. "Transsion Holdings." www.transsion.com/en/?language=en.

China Daily. "China's Transsion Phones Outsell Samsung in Africa." September 2, 2017. www.chinadaily.com.cn/business/tech/2017-02/09/content_28144012.htm.

Lonely Planet. "Lonely Planet Zambia, Mozambique & Malawi." September 19, 2017.

South China Morning Post. "How Transsion Became No 3 in India by Solving Oily Fingers Problem." January 12, 2018. www.scmp.com/tech/social-gadgets/article/2127671/how-unknown-chinese-phone-maker-became-no-3-india-solving-oily.

Internet.org. "Internet.org." info.internet.org/en/.

Internet.org. "Free Basics by Facebook—English." October 12, 2016. info.internet.org/en/story/free-basics-from-internet-org/.

Internet.org. "Reach the Next Wave of People Coming to the Internet." October 12, 2016. info.internet.org/en/story/platform/.

Wikipedia. s.v. "Facebook Aquila." Accessed July 3, 2018. en.wikipedia.org/wiki/Facebook_Aquila.

X. "Project Loon." x.company/loon/.

Mirani, Leo. "Millions of Facebook Users Have No Idea They're Using the Internet." *Quartz*, February 9, 2015. qz.com/333313/milliions-of-facebook-users-have-no-idea-theyre-using-the-Internet/.

Constine, Josh. "Facebook Now Has 2 Billion Monthly Users . . . and Responsibility." TechCrunch, June 27, 2017. techcrunch.com/2017/06/27/facebook-2-billion-users/.

POSSIBILITY

Gorongosa National Park. "Welcome." www.gorongosa.org/.

E. O. Wilson Biodiversity Foundation. "E. O. Wilson Biodiversity Laboratory at Gorongosa National Park." eowilsonfoundation.org/e-o-wilson-laboratory-at-gorongosa/.

Wikipedia. s.v. "Mozambican Civil War." Accessed July 4, 2018. en.wikipedia.org/wiki/Mozambican_Civil_War.

Wikipedia. s.v. "Gregory C. Carr." Accessed July 7, 2018. en.wikipedia.org/wiki/Gregory_C._Carr.

Smithsonian. "Greg Carr's Big Gamble." May 1, 2007. www.smithsonianmag.com/travel/greg-carrs-big-gamble-153081070/.

Deutsche Welle. "Literally Bringing Life Back to Mozambique's Gorongosa National Park." DW.com, October 11, 2017. www.dw.com/en/literally-bringing-life-back-to-mozambiques-gorongosa-national-park/a-40968419.

Steele, John. "Ingenious: Greg Carr—Issue 28: 2050." *Nautilus*, September 17, 2015. nautil.us/issue/28/2050/ingenious-greg-carr.

Jhunjhunwala, Ashok. "Innovative Direct-Current Microgrids to Solve India's Power Woes." *IEEE Spectrum*, January 31, 2017. spectrum.ieee.org/energy/renewables/innovative-direct-current-microgrids-to-solve-indias-power-woes.

Saubhagya Dashboard. saubhagya.gov.in/dashboard.

Ministry of Power. "Rural Electrification." powermin.nic.in/en/content/overview-1.

World Health Organization. "Household Air Pollution and Health." www.who.int/news-room/fact-sheets/detail/household-air-pollution-and-health.

Global Alliance for Clean Cook Stoves. "About." cleancookstoves.org/about/.

Economic Times. "DBT for LPG Is World's Largest Direct Benefit Transfer Scheme." February 5, 2015. economictimes.indiatimes.com/news/politics-and-nation/dbt-for-lpg-is-worlds-largest-direct-benefit-transfer-scheme/articleshow/46131585.cms.

PTI. "India's LPG Cash Subsidy in Bank a/c among Largest Cash Transfer Schemes in World." *Times of India*, February 5, 2015. timesofindia.indiatimes.com/business/india-business/Indias-LPG-cash-subsidy-in-bank-a/c-among-largest-cash-transfer-schemes-in-world/articleshow/46131775.cms.

Wikipedia. s.v. "Optical Fiber." Accessed July 7, 2018. en.wikipedia.org/wiki/Optical_fiber.

Wikipedia. s.v. "Submarine Communications Cable." Accessed July 7, 2018. en.wikipedia.org/wiki/Submarine_communications_cable.

Tripp Lite. "7 Advantages of Fiber Optic Cables over Copper Cables." *Tripp Lite* (blog), September 26, 2016. blog.tripplite.com/7-advantages-of-fiber-optic-cables-over-copper-cables/.

Gonzalez, Sarah, and Karen Duffin. "Episode 853: Peak Sand." NPR, July 13, 2018. www.npr.org/sections/money/2018/07/13/628894815/episode-853-peak-sand?utm_source=npr_newsletter&utm_medium=email&utm_content=20180714&utm_campaign=money&utm_term=nprnews.

Wikipedia. s.v. "TAT-8." Accessed July 3, 2018. en.wikipedia.org/wiki/TAT-8.

Many Possibilities. "African Undersea Cables." May 15, 2018. manypossibilities.net/african-undersea-cables/.

AfricaNews. "Somalia Hit by Internet Outage after Fibre Optic Cables Are Cut by Ship." June 28, 2017. www.africanews.com/2017/06/29/somalia-hit-by-internet-outage-after-fibre-optic-cables-are-cut-by-ship/.

Liquid Telecom. "Network Map." www.liquidtelecom.com/about-us/network-map.html.

Agency Staff. "Google Is Laying Fibre Optic Cable in Africa to Ease Access to the Internet." *Business Day*, March 15, 2017. www.businesslive.co.za/bd/

companies/2017-03-15-google-is-laying-fibre-optic-cable-in-africa-to-ease-access-to-the-internet/.

Economist. "Beefing up Mobile-Phone and Internet Penetration in Africa." November 9, 2017. www.economist.com/special-report/2017/11/09/beefing-up-mobile-phone-and-internet-penetration-in-africa.

Wikipedia. s.v. "Internet Exchange Point." Accessed July 3, 2018. en.wikipedia.org/wiki/Internet_exchange_point.

Hamels Foundation. "Home." www.thehamelsfoundation.org/.

Wikipedia. s.v. "Cole Hamels." Accessed July 8, 2018. en.wikipedia.org/wiki/Cole_Hamels.

Survivor Wiki. "Heidi Strobel." survivor.wikia.com/wiki/Heidi_Strobel.

United Nations. "United Nations Millennium Development Goals." www.un.org/millenniumgoals/.

Collins English Dictionary. s.v. "Welcome Wagon." https://www.collinsdictionary.com/us/dictionary/english/welcome-wagon.

Helen Diller Family Comprehensive Cancer Center. "Jay A. Levy, MD." cancer.ucsf.edu/people/profiles/levy_jay.3417.

Phiri, Frank. "Malawi Can Eradicate HIV Infections Says U.S. Doctor Who Discovered AIDS Virus." Reuters, April 18, 2018. www.reuters.com/article/us-malawi-health-aids/malawi-can-eradicate-hiv-infections-says-u-s-doctor-who-discovered-aids-virus-idUSKBN1HP2HC.

INVESTMENT

Hammond, Carl F. "Buckaroo Hall of Fame." Buckaroo Hall of Fame. www.buckaroohalloffame.com/.

Wikipedia. s.v. "Miss Atomic (Pageants)." Accessed July 3, 2018. en.wikipedia.org/wiki/Miss_Atomic_(pageants).

Burning Man. "Welcome Home." burningman.org/.

Long, Matt. "Driving the Extraterrestrial Highway in Nevada: What You Need to Know." LandLopers, September 5, 2016. landlopers.com/2016/09/13/extraterrestrial-highway.

X. "FSOC at X." x.company/projects/fsoc/.

Terdiman, Daniel. "Alphabet Uses AI to Expand Coverage for Its Internet-Beaming Balloon Network by Orders of Magnitude." *Fast Company,* February 16, 2017. www.fastcompany.com/4030835/google-uses-ai-to-expand-coverage-for-its-internet-beaming-balloon-network-by-orders-of-magnitude.

Locker, Melissa. "Puerto Rico Cell Phone Service Is Still Out. Can Alphabet's Project Loon Fix It?" *Fast Company*, October 9, 2017. www.fastcompany.com/40478905/puerto-rico-cellphone-service-is-still-out-can-alphabets-project-loon-restore-it.

Westgarth, Alastair. "Turning on Project Loon in Puerto Rico—X, the Moonshot Factory." *Loon Blog*, October 20, 2017. blog.x.company/turning-on-project-loon-in-puerto-rico-f3aa41ad2d7f.

Mattise, Nathan. "Project Loon Signs Its First Deal for Internet-Delivering Balloons—in Kenya." *Ars Technica*, July 19, 2018. arstechnica.com/gadgets/2018/07/project-loon-signs-its-first-deal-for-internet-delivering-balloons-in-kenya/.

Zuckerberg, Mark. "The Technology behind Aquila." Facebook, July 22, 2016. www.facebook.com/notes/mark-zuckerberg/the-technology-behind-aquila/10153916136506634/.

Sherr, Ian. "Facebook's Internet-Beaming Drone Is Getting Better." CNET, April 19, 2017. www.cnet.com/news/facebook-f8-acquila-drone-internet-wi-fi-getting-better/.

Glaser, April. "Here's Why Facebook's Massive Drone Crashed in the Arizona Desert." Recode, December 18, 2016. www.recode.net/2016/12/18/13998900/facebooks-drone-crash-aquila-arizona-structural-failure.

Facebook Code. "High Altitude Connectivity: The Next Chapter." June 27, 2018. code.fb.com/connectivity/high-altitude-connectivity-the-next-chapter/.

Airborne Wireless Network. www.airbornewirelessnetwork.com/index.asp.

Irving, Michael. "Will Plans to Use Commercial Aircraft for a Worldwide Wireless Network Fly?" *New Atlas*, December 23, 2016. newatlas.com/airborne-wireless-network-aircraft-internet/47093/.

Irving, Michael. "LandCruisers to Become Roving Communications Hotspots in the Australian Outback." *New Atlas*, May 19, 2016. newatlas.com/landcruisers-outback-roving-communication-hotspots/43356/.

Angier, Natalie. "In Mozambique, a Living Laboratory for Nature's Renewal." *New York Times*, July 23, 2018. www.nytimes.com/2018/07/23/science/gorongosa-animals-environment.html.

Russell, Kendall. "SpaceX Picks Name for Its Satellite Broadband Network." Via Satellite, September 22, 2017. www.satellitetoday.com/innovation/2017/09/22/spacex-picks-name-satellite-broadband-network/.

Snider, Mike. "SpaceX Gets Closer to Launching Satellite Broadband Internet Service." *USA Today*, February 16, 2018. www.usatoday.com/story/tech/news/2018/02/14/fcc-chairman-oks-spacex-bid-deliver-satellite-broadband-internet-service/337283002/.

Russell, Kendall. "SpaceX Testifies: First Prototype Satellite Coming This Year." Via Satellite, November 2, 2017. www.satellitetoday.com//telecom/2017/11/02/spacex-testifies-first-prototype-satellite-coming-year/.

Ralph, Eric, and Taylor S. Marks. "SpaceX Could Provide Satellite Broadband Service as Early as 2020." Teslarati.com, October 27, 2017. www.teslarati.com/spacex-starlink-satellite-broadband-internet-service-2020/.

Hackwill, Robert. "India Claims Record for Multiple Satellite Launch." Euronews, February 15, 2017. www.euronews.com/2017/02/15/india-claims-record-for-multiple-satellite-launch.

SpaceX. "Open Positions." www.spacex.com/careers/list?location[]=906.

Winkler, Rolfe, and Andy Pasztor. "Exclusive Peek at SpaceX Data Shows Loss in 2015, Heavy Expectations for Nascent Internet Service." *Wall Street Journal*, January 13, 2017. www.wsj.com/articles/exclusive-peek-at-spacex-data-shows-loss-in-2015-heavy-expectations-for-nascent-internet-service-1484316455.

Mercury News. "SpaceX Gets OK to Launch High-Speed Satellite Internet Service." March 30, 2018. www.mercurynews.com/2018/03/29/spacex-gets-ok-to-launch-high-speed-satellite-internet-service/.

NASASpaceFlight.com. "SpaceX's Mr. Steven, the FSV Fairing Catcher." www.nasaspaceflight.com/2018/02/spacexs-mr-steven-fsv-fairing-catcher/.

Bonnington, Christina. "The Story behind 'Mr. Steven,' SpaceX's Fairing-Catcher Boat." *Daily Dot*, February 22, 2018. www.dailydot.com/debug/mr-steven-fairing-catcher-boat/.

Space.com. "Meet Mr. Steven, SpaceX's Rocket Nose-Cone-Catching Boat." https://www.space.com/41614-spacex-mr-steven-catcher-boat-up-close.html.

Wikipedia. s.v. "OneWeb Satellite Constellation." Accessed July 9, 2018. en.wikipedia.org/wiki/OneWeb_satellite_constellation.

LeoSat. "LeoSat Enterprises: A New Satellite Paradigm and Unique Data Network Solution." leosat.com/.

Staff Writer. "Boeing Open to Partnerships on LEO Broadband Constellation." Via Satellite, September 21, 2016. www.satellitetoday.com/telecom/2016/09/20/boeing-open-partnerships-leo-broadband-constellation/.

SES. "Beyond Frontiers." July 3, 2018. www.ses.com/.

Viasat. "Inspired to Connect the World." viasat.com/.

Science and Technology Daily. "The SpaceX Starlink Program 'China Edition' Is Coming." March 5, 2018. http://www.stdaily.com/zhuanti01/hangkong/2018-03/05/content_643752.shtml.

Irving, Michael. "Global Quantum Internet Dawns, Thanks to China's Micius Satellite." *New Atlas*, January 24, 2018. newatlas.com/micius-quantum-internet-encryption/53102/.

Cheng, Kelsey. "Chinese Tech Firm Unveils the First Satellite in Its Ambitious Plan to Provide Free Worldwide Wi-Fi." *Daily Mail Online*, November 29, 2018. www.dailymail.co.uk/news/article-6442441/Chinese-tech-firm-unveils-satellite-ambitious-plan-provide-free-worldwide-Wi-Fi.html.

GBTIMES. "Chinese Commercial Rocket Company Secures 1.2 Billion Yuan Investment, Multiple Launches Set for 2018." December 19, 2017. gbtimes.com/chinese-commercial-rocket-company-secures-12bn-yuan-investment-multiple-launches-set-for-2018.

Magan, Veronica. "Could Facebook Be Building Its Own Satellite?" Via Satellite, May 3, 2018. www.satellitetoday.com/telecom/2018/05/03/could-facebook-be-building-its-own-satellite/.

Wikipedia. s.v. "Phased Array." Accessed July 10, 2018. en.wikipedia.org/wiki/Phased_array.

Business Wire. "LeoSat and Phasor Reach Strategic Agreement to Bring Game-Changing Connectivity to Mission Critical Enterprise Networks." March 13, 2018. www.businesswire.com/news/home/20180313005516/en/LeoSat-Phasor-Reach-Strategic-Agreement-Bring-Game-Changing.

Business Wire. "SES Networks Announces Partnerships for Groundbreaking O3b MPOWER Customer Edge Terminals." March 8, 2018. www.businesswire.com/news/home/20180307006505/en/SES-Networks-Announces-Partnerships-Groundbreaking-O3b-mPOWER.

Stratolaunch. www.stratolaunch.com/.

Davenport, Christian. "Why Is Paul Allen Building the World's Largest Airplane? Perhaps to Launch a Space Shuttle Called Black Ice." *Washington Post*, March 6, 2018. www.washingtonpost.com/news/the-switch/wp/2018/03/06/why-is-paul-allen-building-the-worlds-largest-airplane-perhaps-to-launch-a-space-shuttle-called-black-ice/?utm_term=.25c3cc56b453.

NBCNews.com. "World's Biggest Plane, Stratolaunch, Marks Another Key Milestone." www.nbcnews.com/mach/science/world-s-biggest-plane-stratolaunch-marks-another-key-milestone-ncna851556.

Levy, Steven. "385 Feet of Crazy: The Most Audacious Flying Machine Ever." *Wired*, August 20, 2018. www.wired.com/story/stratolaunch-airplane-burt-rutan-paul-allen/.

Boyle, Alan. "Paul Allen's Stratolaunch Unveils Plans for New Family of Rockets (and a Space Plane)." *GeekWire*, August 21, 2018. www.geekwire.com/2018/paul-allens-stratolaunch-unveils-plans-new-family-rockets-space-plane/.

Blue Origin. www.blueorigin.com/.

SpaceNews.com. "Blue Origin Still Planning Commercial Suborbital Flights in 2018." April 6, 2017. www.spacenews.com/blue-origin-still-planning-commercial-suborbital-flights-in-2018/.

Virgin Orbit. "Home." virginorbit.com/.

Wikipedia. s.v. "LauncherOne." Accessed July 9, 2018. en.wikipedia.org/wiki/LauncherOne.

Rocket Lab. "Space Is Now Open for Business." www.rocketlabusa.com/.

CubeSat. "Home." www.cubesat.org/.

Bloomberg. "This Startup Got $40 Million to Build a Space Catapult." June 14, 2018. www.bloomberg.com/news/articles/2018-06-14/this-startup-got-40-million-to-build-a-space-catapult.

Boyle, Alan. "SpinLaunch Raises $40M from Airbus, Google and Others for Space Catapult." *GeekWire*, June 15, 2018. www.geekwire.com/2018/spinlaunch-raises-40m-airbus-google-others-space-catapult/.

Loff, Sarah. "CubeSats Overview." NASA, July 22, 2015. www.nasa.gov/mission_pages/cubesats/overview.

Jackson, Shanessa. "NASA's CubeSat Launch Initiative." NASA, February 17, 2017. www.nasa.gov/directorates/heo/home/CubeSats_initiative.

Wikipedia. s.v. "CubeSat." Accessed July 10, 2018. en.wikipedia.org/wiki/CubeSat.

Howell, Elizabeth. "CubeSats: Tiny Payloads, Huge Benefits for Space Research." Space.com, June 19, 2018. www.space.com/34324-cubesats.html.

Sky and Space Global. "Sky and Space Global." www.skyandspace.global/.

Konnect Africa. "Konnect Africa—Provider of Satellite Broadband Solutions in Africa." www.konnect-africa.com/.

Konnect Africa. "SMART WiFi | Konnect." www.konnect-africa.com/smart-wifi.

NASASpaceFlight.com. "Bangabandhu-1 Successfully Launched by First Block 5 Falcon 9—SpaceX's Goal of Affordable Access to Space." www.nasaspaceflight.com/2018/05/bangabandhu-1-launch-spacexs-affordable-space/.

Economist. "What Technology Can Do for Africa." November 9, 2017. www.economist.com/special-report/2017/11/09/what-technology-can-do-for-africa.

M-KOPA Solar. www.m-kopa.com/.

Coldewey, Devin. "FCC Approves SpaceX Plan for 4,425-Satellite Broadband Network." TechCrunch, March 29, 2018. techcrunch.com/2018/03/29/fcc-approves-spacex-plan-for-4425-satellite-broadband-network/.

Wikipedia. s.v. "Kessler Syndrome." Accessed July 6, 2018. en.wikipedia.org/wiki/Kessler_syndrome.

Bill and Melinda Gates Foundation. "Financial Services for the Poor." www.gatesfoundation.org/What-We-Do/Global-Growth-and-Opportunity/Financial-Services-for-the-Poor.

Wikipedia. s.v. "List of Wealthiest Charitable Foundations." Accessed July 5, 2018. en.wikipedia.org/wiki/List_of_wealthiest_charitable_foundations.

Bill and Melinda Gates Foundation. "Bill and Melinda Gates Foundation." www.gatesfoundation.org/.

Bill and Melinda Gates Foundation. "Foundation Fact Sheet." www.gatesfoundation.org/Who-We-Are/General-Information/Foundation-Factsheet.

Bill and Melinda Gates Foundation. "Foundation Trust." www.gatesfoundation.org/Who-We-Are/General-Information/Financials/Foundation-Trust.

GROWTH

ICT4D Conference. www.ict4dconference.org/.

BongoHive. "Home." bongohive.co.zm/.

BYJU'S. "BYJUS e Learning for Online Courses." byjus.com/.

Economist. "Indian Teaching Startups Make Work for Idle Thumbs." February 17, 2018. www.economist.com/business/2018/02/17/indian-teaching-startups-make-work-for-idle-thumbs.

XPRIZE. "Global Learning XPRIZE." www.xprize.org/prizes/global-learning.

Unique Identification Authority of India. "About UIDAI." www.uidai.gov.in/about-uidai/about-uidai.html.

Wikipedia. s.v. "Aadhaar." Accessed July 13, 2018. en.wikipedia.org/wiki/Aadhaar.

MediaNama. "Over 1.21 Billion Enrolled onto Aadhaar, with 3.58 Million in June 2018." July 13, 2018. www.medianama.com/2018/07/223-aadhaar-enrolment-june-2018/.

Element. "Element: Delivering Digital Identity." www.discoverelement.com/.

Zipline International Inc. "Lifesaving Deliveries by Drone." www.flyzipline.com/.

Praekelt.org. "MomConnect." www.praekelt.org/momconnect.

JNJ. "MomConnect: Connecting Women to Care, One Text at a Time." January 31, 2017. www.jnj.com/our-giving/momconnect-connecting-women-to-care-one-text-at-a-time.

Shapshak, Toby. "South African Messaging Wonder MomConnect Launches on WhatsApp." *Forbes,* December 4, 2017. www.forbes.com/sites/

tobyshapshak/2017/12/04/african-messaging-wonder-momconnect-launches-on-whatsapp/#2c970c657c3b.

MPower. "MPower." www.mpower-social.com/.

OpenSRP. "Open Smart Register Platform." smartregister.org/.

GiveDirectly. "GiveDirectly: Send Money Directly to the Extreme Poor." www.givedirectly.org/.

GiveDirectly. "GiveDirectly Frequently Asked Questions." www.givedirectly.org/faq.

Linke, Rebecca. "12-Year Study Looks at Effects of Universal Basic Income." MIT Sloan School of Management, January 30, 2018. mitsloan.mit.edu/newsroom/articles/12-year-study-looks-at-effects-of-universal-basic-income/.

Segovia. "Frontier Market Payments for Global Businesses." www.thesegovia.com/.

Goodman, Peter S. "Capitalism Has a Problem. Is Free Money the Answer?" *New York Times*, November 15, 2017. www.nytimes.com/2017/11/15/business/dealbook/universal-basic-income.html.

Lowrey, Annie. "The Future of Not Working." *New York Times*, February 23, 2017. www.nytimes.com/2017/02/23/magazine/universal-income-global-inequality.html.

Wikipedia. s.v. "M-Pesa." Accessed July 12, 2018. en.wikipedia.org/wiki/M-Pesa.

Vodafone. "M-Pesa from Vodafone." www.vodafone.com/content/index/what/m-pesa.html.

Monks, Kieron. "M-Pesa: Kenya's Mobile Success Story Turns 10." CNN, February 24, 2017. www.cnn.com/2017/02/21/africa/mpesa-10th-anniversary/index.html.

Georgetown University. "Study: Use of Mobile Money Lifts Nearly 200,000 Kenyans Out of Poverty." www.georgetown.edu/Billy-Jack-Mobile-Money-Kenya-research.

What3words. "Addressing the World." what3words.com/.

Nature News. "Satellite Images Reveal Gaps in Global Population Data." www.nature.com/news/satellite-images-reveal-gaps-in-global-population-data-1.21957.

World Bank. "Why Secure Land Rights Matter." www.worldbank.org/en/news/feature/2017/03/24/why-secure-land-rights-matter.

Capital Flows. "Property Rights of the Poor Need to Be Recognized in Developing Countries." *Forbes*, January 8, 2015. www.forbes.com/sites/realspin/2015/01/08/property-rights-of-the-poor-need-to-be-recognized-in-developing-countries/#2dd9a3944cf2.

Wayumba, Robert. "Drones Are Taking to the Skies above Africa to Map Land Ownership." *The Conversation*, July 13, 2018. theconversation.com/drones-are-taking-to-the-skies-above-africa-to-map-land-ownership-87369.

Copenhagen Consensus Center. www.copenhagenconsensus.com/.

Lomborg, Bjorn. "Digital Solutions Can Help Even the Poorest Nations Prosper." *Wired*, October 28, 2017. www.wired.com/story/digital-solutions-can-help-even-the-poorest-nations-prosper/?mc_cid=804581b3e6&mc_eid=d74fc4bde5.

Coca-Cola Company. "Ekocenter." www.coca-colacompany.com/ekocenter.

Russell, Kendall. "Coca-Cola Leverages Intelsat VSATs to Stimulate Rural Growth." Via Satellite, March 30, 2018. www.satellitetoday.com/telecom/2018/03/30/coca-cola-leverages-intelsat-vsats-to-stimulate-rural-growth/.

Coca-Cola Company. "#5by20." www.coca-colacompany.com/5by20.

Phandeeyar. "Our Programs." phandeeyar.org/.

Fortune. "Asia's Least-Developed Telecom Market Will Soon Become the World's Fastest Growing." September 18, 2014. fortune.com/2014/09/18/asia-myanmar-burma-telecommunications-market/.

Wikipedia. s.v. "List of Ethnic Groups in Myanmar." Accessed July 12, 2018. en.wikipedia.org/wiki/List_of_ethnic_groups_in_Myanmar.

Wikipedia. s.v. "Myanmar." Accessed July 11, 2018. en.wikipedia.org/wiki/Myanmar.

CHALLENGE

Lonely Planet. "Cox's Bazar." www.lonelyplanet.com/bangladesh/chittagong-division/coxs-bazar.

Miles, Tom. "U.N. Investigators Cite Facebook Role in Myanmar Crisis." Reuters, March 12, 2018. www.reuters.com/article/us-myanmar-rohingya-facebook/u-n-investigators-cite-facebook-role-in-myanmar-crisis-idUSKCN1GO2PN.

Munita, Tómas, Ben C. Solomon, Mona El-Naggar, and Adam Dean. "How the Rohingya Escaped." *New York Times*, December 21, 2017. www.nytimes.com/interactive/2017/12/21/world/asia/how-the-rohingya-escaped.html.

Clark, Doug Bock. "Myanmar's Internet Disrupted Society—and Fueled Extremists." *Wired*, September 27, 2017. www.wired.com/story/myanmar-internet-disrupted-society-extremism/.

Oppenheim, Marella. "'It Only Takes One Terrorist': The Buddhist Monk Who Reviles Myanmar's Muslims." *Guardian*, May 12, 2017. www.theguardian

.com/global-development/2017/may/12/only-takes-one-terrorist-buddhist-monk-reviles-myanmar-muslims-rohingya-refugees-ashin-wirathu.

TRTWorld. "Anti-Muslim Monk Preaches Hate: Ashin Wirathu." Published on September 8, 2017. YouTube video, 1:50. www.youtube.com/watch?v=16YhQ4JWHYU.

Mozur, Paul. "A Genocide Incited on Facebook, with Posts from Myanmar's Military." *New York Times*, October 15, 2018. www.nytimes.com/2018/10/15/technology/myanmar-facebook-genocide.html.

Facebook Newsroom. "Hard Questions: Q&A with Mark Zuckerberg on Protecting People's Information." April 4, 2018. newsroom.fb.com/news/2018/04/hard-questions-protecting-peoples-information/.

Kleiner Perkins Caufield & Byers. "Internet Trends 2018." LinkedIn SlideShare, May 30, 2018. www.slideshare.net/kleinerperkins/internet-trends-report-2018-99574140.

Airoldi, Donna. "UNHCR Chief Urges Myanmar to Grant Rohingya Citizenship." Reuters, July 7, 2017. www.reuters.com/article/us-asia-refugees/unhcr-chief-urges-myanmar-to-grant-rohingya-citizenship-idUSKBN19S1VJ.

Nyi Nyi, U. "The Case against Rohingya Citizenship." *Myanmar Times*, February 9, 2014. www.mmtimes.com/opinion/9519-why-the-government-is-right-to-deny-rohingya-citizenship.html.

Wikipedia. s.v. "Rohingya People." Accessed July 11, 2018. en.wikipedia.org/wiki/Rohingya_people.

ReliefWeb. "Bangladesh—the Rohingya Refugee Camp of Kutupalong-Balukhali, an Ever-Expanding Camp." reliefweb.int/report/bangladesh/bangladesh-rohingya-refugee-camp-kutupalong-balukhali-ever-expanding-camp.

World Food Program USA. "A Look into the World's Largest Refugee Camp." wfpusa.org/articles/a-firsthand-look-into-the-worlds-largest-refugee-camp/.

Wikipedia. s.v. "Kutupalong Refugee Camp." Accessed July 10, 2018. en.wikipedia.org/wiki/Kutupalong_refugee_camp.

Pakistan Media and Telecoms Landscape Guide. "Information Changes Lives." Internews. www.internews.org/.

International Organization for Migration. "International Organization for Migration." www.iom.int/.

UNHCR. "United Nations." www.unhcr.org/en-us/.

BRAC. "Monsoon Hits Cox's Bazar: The World's Biggest Makeshift City." Published on May 10, 2018. YouTube video, 1:02. www.youtube.com/watch?v=ihc3xHuFZAQ&feature=youtu.be.

Chowdhury, Zaian. "9 Ways We Are Keeping People in Cox's Bazar Safe in Monsoon." *BRAC Blog*, June 28, 2018. blog.brac.net/9-ways-we-are-keeping-people-in-coxs-bazar-safe-in-monsoon/.

Friedman, Thomas L. "Alabama Says No to Trump's Tribalism." *New York Times*, December 13, 2017. www.nytimes.com/2017/12/13/opinion/alabama-senate-election-jones.html.

Rashid, Tania. "PBS Newshour, Rohingya." Public Broadcasting Service. www.pbs.org/newshour/author/tania-rashid.

Goel, Vindu, et al. "In Sri Lanka, Facebook Contends with Shutdown after Mob Violence." *New York Times*, March 8, 2018. www.nytimes.com/2018/03/08/technology/sri-lanka-facebook-shutdown.html.

Taub, Amanda, and Max Fisher. "Where Countries Are Tinderboxes and Facebook Is a Match." *New York Times*, April 21, 2018. www.nytimes.com/2018/04/21/world/asia/facebook-sri-lanka-riots.html.

Dwoskin, Elizabeth, and Annie Gowen. "On WhatsApp, Fake News Is Fast—and Can Be Fatal." *Washington Post*, July 23, 2018. https://www.washingtonpost.com/business/economy/on-whatsapp-fake-news-is-fast--and-can-be-fatal/2018/07/23/a2dd7112-8ebf-11e8-bcd5-9d911c784c38_story.html?utm_term=.102fbec160d4.

U.S. Embassy and U.S. Citizens. "Security Message for U.S. Citizens: Attacks on and around Mt. Mulanje." U.S. Embassy in Malawi, September 19, 2017. mw.usembassy.gov/security-message-u-s-citizens-attacks-around-mt-mulanje/.

Maravi Post. "Malawi Legislature Admits the Existence of Bloodsuckers." October 10, 2017. www.maravipost.com/malawi-legislature-admits-existence-bloodsuckers/.

Allison, Simon. "Vigilante Vampire Slayers Haunt Malawi." *M&G Online*, October 20, 2017. mg.co.za/article/2017-10-20-00-vigilante-vampire-slayers-haunt-malawi.

Editorial Board. "The 'Deep Fake' Threat." Bloomberg.com, June 13, 2018. www.bloomberg.com/view/articles/2018-06-13/the-deep-fake-video-threat.

Hui, Jonathan. "How Deep Learning Fakes Videos (Deepfakes) and How to Detect It." Medium, April 28, 2018. medium.com/@jonathan_hui/how-deep-learning-fakes-videos-deepfakes-and-how-to-detect-it-c0b50fbf7cb9.

New Statesman. "How to Identify If an Online Video Is Fake." February 14, 2018. www.newstatesman.com/science-tech/technology/2018/02/how-identify-if-online-video-fake.

BRAC. "BRAC." www.brac.net.

Kirby, Jen. "Zuckerberg: Facebook Has Systems to Stop Hate Speech. Myanmar Groups: No, It Doesn't." Vox, April 6, 2018. www.vox.com/2018/4/6/

17204324/zuckerberg-facebook-myanmar-rohingya-hate-speech-open-letter.

Roose, Kevin, and Paul Mozur. "Zuckerberg Was Called Out over Myanmar Violence. Here's His Apology." *New York Times*, April 9, 2018. www.nytimes.com/2018/04/09/business/facebook-myanmar-zuckerberg.html.

Dellinger, A. J. "Mark Zuckerberg Emails His Myanmar Critics Directly, They Publicly Blast Back." Gizmodo, April 10, 2018. gizmodo.com/mark-zuckerberg-emails-his-myanmar-critics-directly-th-1825129737.

Facebook. "Terms of Service." www.facebook.com/terms.php.

Scherker, Amanda. "Didn't Read Facebook's Fine Print? Here's Exactly What It Says." *Huffington Post*, December 7, 2017. www.huffingtonpost.com/2014/07/21/facebook-terms-condition_n_5551965.html.

Oltermann, Philip. "Tough New German Law Puts Tech Firms and Free Speech in Spotlight." *Guardian*, January 5, 2018. www.theguardian.com/world/2018/jan/05/tough-new-german-law-puts-tech-firms-and-free-speech-in-spotlight.

DPA/The Local. "Facebook to Hire 500 Workers in Essen to Delete Hate Speech." *The Local*, August 9, 2017. www.thelocal.de/20170809/facebook-to-hire-500-employees-in-essen-to-combat-hate-speech-socialmedia.

Balakrishnan, Anita. "Facebook Pledges to Double Its 10,000-Person Safety and Security Staff by End of 2018." CNBC, November 1, 2017. www.cnbc.com/2017/10/31/facebook-senate-testimony-doubling-security-group-to-20000-in-2018.html.

ArchDaily. "Facebook Moves into New Headquarters with the 'Largest Open Floor Plan in the World.'" March 30, 2015. www.archdaily.com/614515/facebook-moves-into-new-headquarters-with-the-largest-open-floor-plan-in-the-world.

Facebook Newsroom. "Update on Myanmar." newsroom.fb.com/news/2018/08/update-on-myanmar/.

Coldewey, Devin. "Facebook Independent Research Commission, Social Science One, Will Share a Petabyte of User Interactions." TechCrunch, July 11, 2018. techcrunch.com/2018/07/11/facebook-independent-research-commission-social-science-one-will-share-a-petabyte-of-user-data/.

Special Reports. "Why Facebook Is Losing the War on Hate Speech in Myanmar." Reuters, August 15, 2018. www.reuters.com/investigates/special-report/myanmar-facebook-hate/.

Stewart, Emily. "The $120-Billion Reason We Can't Expect Facebook to Police Itself." Vox, July 28, 2018. www.vox.com/business-and-finance/2018/7/28/17625218/facebook-stock-price-twitter-earnings.

Pender, Kathleen. "Yes, Median Pay at Facebook Really Is about $240,000 a Year." *San Francisco Chronicle*, April 29, 2018. www.sfchronicle.com/business/networth/article/Yes-median-pay-at-Facebook-really-is-about-12870786.php.

Pakistan Media and Telecoms Landscape Guide. "Can We Solve America's Fake News Problem? A Media Expert's Advice." Internews. www.internews.org/news/can-we-solve-americas-fake-news-problem-media-experts-advice.

Chokshi, Niraj. "How to Fight 'Fake News' (Warning: It Isn't Easy)." *New York Times*, September 18, 2017. www.nytimes.com/2017/09/18/business/media/fight-fake-news.html.

Wendling, Mike. "Solutions That Can Stop Fake News Spreading." BBC News, January 30, 2017. www.bbc.com/news/blogs-trending-38769996.

West, Darrell M. "How to Combat Fake News and Disinformation." Brookings, December 18, 2017. www.brookings.edu/research/how-to-combat-fake-news-and-disinformation/.

Economist. "On Twitter, Falsehood Spreads Faster Than Truth." March 10, 2018. www.economist.com/science-and-technology/2018/03/10/on-twitter-falsehood-spreads-faster-than-truth.

Nagarajah, Sasha. "Zawgyi vs. Unicode." *Global App Testing* (blog). www.globalapptesting.com/blog/zawgyi-vs-unicode.

Bloomberg. "The Great Firewall of China." October 13, 2017. www.bloomberg.com/quicktake/great-firewall-of-china.

Wikipedia. s.v. "Websites Blocked in Mainland China." Accessed July 5, 2018. en.wikipedia.org/wiki/Websites_blocked_in_mainland_China.

Wong, Edward. "Riots in Western China amid Ethnic Tension." *New York Times*, July 5, 2009. www.nytimes.com/2009/07/06/world/asia/06china.html.

Economist. "China Has Turned Xinjiang into a Police State like No Other." May 31, 2018. www.economist.com/briefing/2018/05/31/china-has-turned-xinjiang-into-a-police-state-like-no-other.

Anderson, Lora R. "Mark Zuckerberg Speaks Mandarin in Chinese New Year." Published on February 7, 2016. YouTube video, 1:24. www.youtube.com/watch?v=fISvHRJWHPg.

New York Times. "Facebook Faces a New World as Officials Rein In a Wild Web." September 18, 2017. www.nytimes.com/2017/09/17/technology/facebook-government-regulations.html.

ITA National Travel and Tourism Office. "Monthly Departures to International Destinations." travel.trade.gov/view/m-2017-O-001/index.html.

PARTNERSHIP

Diamandis, Peter H. *Abundance: The Future Is Better Than You Think.* New York: Simon & Schuster, 2015.

Economist. "The Meaning of the Vision Fund." May 12, 2018. www.economist.com/leaders/2018/05/12/the-meaning-of-the-vision-fund.

NGO Advisor. "Top 100 NGOs." www.ngoadvisor.net/top100ngos/.

PR Newswire. "Deloitte: Americans Look at Their Smartphones More Than 12 Billion Times Daily, Even as Usage Habits Mature and Device Growth Plateaus." November 15, 2017. www.prnewswire.com/news-releases/deloitte-americans-look-at-their-smartphones-more-than-12-billion-times-daily-even-as-usage-habits-mature-and-device-growth-plateaus-300555703.html.

Government of Dubai. "Dubai History." www.dubai.ae/en/aboutdubai/Pages/DubaiHistory.aspx.

Statista. "Facebook Users Reach by Device 2018." www.statista.com/statistics/377808/distribution-of-facebook-users-by-device/.

Jio. "The World's Largest Data Network Based on Mobile Data Consumption." www.jio.com/.

Purnell, Newley. "Two Years Ago, India Lacked Fast, Cheap Internet—One Billionaire Changed All That." *Wall Street Journal,* September 5, 2018. www.wsj.com/articles/two-years-ago-india-lacked-fast-cheap-internetone-billionaire-changed-all-that-1536159916?ns=prod/accounts-wsj.

Rosenberg, Mike, and Ángel González. "Thanks to Amazon, Seattle Is Now America's Biggest Company Town." *Seattle Times,* November 30, 2017. www.seattletimes.com/business/amazon/thanks-to-amazon-seattle-is-now-americas-biggest-company-town/.

Hilburg, Jonathan. "Amazon's Seattle Spheres Are Set for Public Opening." *Architect's Newspaper,* January 26, 2018. archpaper.com/2018/01/amazon-seattle-biospheres-opening/.

Roser, Max, and Esteban Ortiz-Ospina. "Literacy." Our World in Data. www.ourworldindata.org/literacy.

Amazon Developer. "Build Skills for the Greater Good: Announcing the Next Alexa Skills Challenge." developer.amazon.com/blogs/alexa/post/4e10fa0a-6fb6-4f43-b13b-4729ac1a1430/enter-the-alexa-skills-challenge-tech-for-good.

Amazon Developer. "Announcing the Winners of the Alexa Skills Challenge: Tech for Good." developer.amazon.com/blogs/alexa/post/ca34b954-1c5d-4a59-b326-f45c8df7c89c/alexa-skill-tech-for-good-challenge-winners.

Amazon Developer. "Alexa Skills Kit—Build for Voice with Amazon." developer.amazon.com/alexa-skills-kit.

Amazon Developer. "Alexa Voice Service." developer.amazon.com/alexa-voice-service.

Koetsier, John. "AI Assistants Ranked: Google's Smartest, Alexa's Catching Up, Cortana Surprises, Siri Falls Behind." *Forbes*, April 25, 2018. www.forbes.com/sites/johnkoetsier/2018/04/24/ai-assistants-ranked-googles-smartest-alexas-catching-up-cortana-surprises-siri-falls-behind/#6b1de947492a.

Hassan, Mehedi. "Google Assistant Takes the Crown, Cortana Sits Last on Smart Speaker IQ Test." Thurrott.com, December 21, 2018. www.thurrott.com/google/195724/google-assistant-takes-the-crown-cortana-sits-last-on-smart-speaker-iq-test.

Neuralink. www.neuralink.com/.

Rai, Saritha. "Trying to Speak India's Language(s)." *Bloomberg Businessweek*, November 13, 2017. www.magzter.com/article/Business/Bloomberg-Businessweek/Trying-to-Speak-Indias-Languages.

Weinberger, Matt. "Amazon Has a Master Plan to Turn Alexa into a Full-Fledged Economy All on Its Own." *Business Insider*, January 5, 2018. www.businessinsider.com/how-amazon-alexa-will-create-an-app-store-boom-2018-1?IR=T.

Tesla, Inc. "Tesla 2018 Annual Shareholder Meeting." June 5, 2018. www.tesla.com/shareholdermeeting.

Bary, Emily. "'OK Google, Give Everybody in America a Free Speaker.'" MarketWatch, July 1, 2018. www.marketwatch.com/story/ok-google-give-everybody-in-america-a-free-speaker-2018-06-28.

Boyle, Alan. "SpaceX President Gwynne Shotwell Sees Satellites as Bigger Market Than Rockets." *GeekWire*, May 24, 2018. www.geekwire.com/2018/spacex-president-gwynne-shotwell-sees-satellites-bigger-market-rockets/.

Computer History Museum. "Hall of Fellows." www.computerhistory.org/fellowawards/hall/.

AFTERWORD

Ferris, Dacia J. "High-Altitude Balloons to Provide Cell and Internet Coverage over Kenya." Teslarati.com, January 7, 2019. www.teslarati.com/high-altitude-balloons-cell-internet-coverage-kenya/.

XinhuaNet. "China Launches Tech Experimental Satellite for Wider-Reaching Broadband Connection." December 22, 2018. www.xinhuanet.com/english/2018-12/22/c_137692006.htm.

Brodkin, Jon. "After Delays, OneWeb Launches Its First Six Low-Earth Broadband Satellites." *Ars Technica*, February 28, 2019. arstechnica.com/information-technology/2019/02/oneweb-launches-six-low-earth-satellites-pledges-global-broadband-in-2021/.

Forrester, Chris. "Musk Raises Another $1bn for Mars 'Starship.'" *Digital Media Delivery*. advanced-television.com/2019/01/07/musk-raises-another-1bn-for-mars-starship/.

Boyle, Alan. "SpaceX Seeks FCC Approval for up to 1M Starlink Satellite Earth Stations." *GeekWire*, February 10, 2019. www.geekwire.com/2019/spacex-fcc-starlink-million-earth-stations/.

Economist. "Mukesh Ambani Wants to Be India's First Internet Tycoon." January 26, 2019. www.economist.com/business/2019/01/26/mukesh-ambani-wants-to-be-indias-first-internet-tycoon.

Ralph, Eric. "SpaceX Reveals Falcon Fairing Recovery Progress as Mr. Steven Barely Misses Catch." Teslarati.com, January 8, 2019. www.teslarati.com/spacex-falcon-fairing-recovery-progress-reveal-mr-steven-near-miss/.

TECHNICAL GLOSSARY

2G: A digital mobile communications standard (second generation) supporting voice calls and limited data transmission, such as texting.

3G: A digital mobile communications standard (third generation) supporting wireless access to the internet.

4G: A digital mobile communications standard (fourth generation) allowing wireless access to the internet, but at much higher speeds than 3G. Long-Term Evolution (LTE) is a variant of 4G.

5G: A next-generation digital mobile communications standard (fifth generation) allowing much higher data transmission speeds than 4G.

Amazon Web Services (AWS): A cloud services provider offering computing power, database storage, database management, content delivery, and other services.

Android: An open-source operating system created by Google used for smartphones and tablet computers.

ARPANET: Advanced Research Projects Agency Network, an early packet-switching network implementing the TCP/IP protocol as an early precursor to the internet.

artificial intelligence (AI): A term describing computer systems able to mimic human skills in tasks such as speech recognition, visual perception, and translation between languages.

augmented reality (AR): A set of technologies that superimpose computer-generated images on a user's view, thus creating a composite view of the world.

backhaul: The portion of a network providing linkages between a core network and subnetworks. Backhaul services are generally charged at wholesale commercial rates.

bandwidth: The set of frequencies within a given band, especially those used for transmitting a signal. On a network, bandwidth can also refer to the data transfer rate.

baud: A unit of transmission speed that equals the number of times a signal changes per second. One baud represents one bit per second.

broadband: A high-capacity data-transfer technique employing a wide range of frequencies. This allows a large number of messages to simultaneously be communicated. In the context of internet access, broadband refers to any high-speed access that is persistent and faster than traditional dial-up access.

cell phone: A portable, typically cordless phone for voice communications in a cellular system. Short for cellular phone.

cloud: A term that refers to access of computer and software applications, often through data centers using wide area networking (WAN) or internet connectivity.

connectivity: Capacity for the interconnection of platforms, systems, and applications. In the context of internet access, it refers to connection or communication with other internet-enabled systems.

CubeSat: A miniature satellite made up of multiple 10 × 10 × 10 centimeter units, each weighing no more than 1.33 kilograms, typically employing commercial off-the-shelf (COTS) components for electronics and structure.

deepfake: A technology used to produce or alter video to present something that didn't occur. The word, which refers to both the technologies and the videos created, is a combination of "deep learning" and "fake."

Digital Evolution Index (DEI): A ranking, developed by Tufts University and Mastercard, of the digital development of countries.

digital subscriber line (DSL): A high-speed connection to the internet over a telephone line.

drone: An unmanned ship or aircraft guided by onboard computers or remote control.

E band: The radio frequency range from 60 GHz to 90 GHz in the electromagnetic spectrum.

fairing: An external metal or plastic structure added to increase streamlining and reduce drag, such as on the top of a rocket.

feature phone: A mobile phone that allows voice communications as well as basic features such as the ability to store and play music but without the advanced functionality of a smartphone.

fiber optic: A term relating to thin flexible fibers with a glass core through which light signals can be sent with very little loss of strength.

file transfer protocol (FTP): A protocol that allows file transfer via TCP/IP networks. It is used to transfer files from client to server (upload), from server to client (download), or between two servers.

gantry: A bridgelike structure that supports equipment such as lights, cameras, or a crane.

geographic information system (GIS): A system that captures, stores, analyzes, and presents geographical data.

geosynchronous orbit (GSO): A circular orbit 22,236 miles above the earth's equator with an orbital period matching the earth's rotation. Satellites in geosynchronous orbit appear to be at a fixed position in the sky relative to ground observers.

gigabit (Gb): A unit of information equal to one billion (technically 2^{30}) bits.

Global Connectivity Index (GCI): An index and ranking developed by Huawei of countries' relative connectivity to the internet.

global positioning service (GPS): A satellite navigation system allowing users at sea, on land, and in the air to determine their exact location, velocity, and time anywhere in the world.

Gopher: An early internet-based system developed by the University of Minnesota that allowed the storage and retrieval of information.

Great Firewall of China: The combination of technologies and legislative actions enforced in China to regulate the domestic internet.

gross domestic product (GDP): The value of all goods and services produced within a country in a specific time period.

hotspot: A wireless access point, often in a public location, providing internet access to mobile devices.

Hyperloop: A high-speed ground transportation system proposed by Elon Musk in 2013.

Hypertext Transfer Protocol (HTTP): Rules for transferring files (text, sound, video, graphic images, and other files) on the World Wide Web.

information and communications technologies (ICT): The integration of telecommunications (telephone lines and wireless signals), computers, and software to enable users to access, transmit, store, and manipulate information.

internet exchange point (IXP): The physical infrastructure through which internet service providers (ISPs) and content delivery networks (CDNs) exchange internet traffic between their networks.

Internet of Things (IoT): A system of computing devices, machines, or objects that are provided with unique identifiers (UIDs) and the ability to transfer data over a network.

internet service provider (ISP): A company that provides businesses and individuals access to the internet.

internet: A global computer network providing information and communication facilities consisting of interconnected networks employing standardized communication protocols.

iOS: an operating system used for mobile devices manufactured by Apple Inc.

K_a band: A portion of the microwave part of the electromagnetic spectrum defined as frequencies in the range of 26.5–40 gigahertz.

Kessler syndrome: A scenario (also called the Kessler effect) in which the density of objects in low earth orbit (LEO) is sufficient that collisions between objects cause a cascade where space debris increases the likelihood of further collisions.

K_u band: A portion of the microwave part of the electromagnetic spectrum defined as frequencies in the range of 12–18 gigahertz.

latency: The delay before a transfer of data begins following an instruction for its transfer.

light-emitting diode (LED): A semiconductor device that emits light when an electric current passes through it.

Long-Term Evolution (LTE): A digital mobile communications standard allowing wireless access to the internet. A variant of 4G.

low earth orbit (LEO): A circular earth orbit generally from 400 to 1,000 miles above the earth's surface.

medium earth orbit (MEO): An earth orbit generally from 1,000 to 22,236 miles above the earth's surface, which is between a low earth orbit (LEO) and geosynchronous orbit (GSO).

megabaud (MBd): A data transmission speed of one million bits per second. Equivalent to one megabit per second (Mbps).

megabit (Mb): One million bits of information.

mesh network: A local network topology supporting efficient routing of data. Nodes connect directly and dynamically to many other nodes.

microgrid: A local solar array serving a small network of electricity users that is able to function independently.

microsatellite: A class of small satellite generally weighing between 10 and 100 kilograms.

microwave: An electromagnetic wave in the range of 0.001–0.3 meters, shorter than a normal radio wave but longer than infrared waves. Microwaves are used in communications, radar, and various industrial processes.

Molniya orbit: A highly elliptical and inclined orbit used for satellite communications over high-latitude regions.

Mosaic: An early web browser credited with popularizing the World Wide Web and the internet.

Netscape: An early commercial web browser developed and marketed by the team that introduced the Mosaic browser.

network news transfer protocol (NNTP): A protocol for transporting Usenet news articles among news servers and for posting and reading by end-user applications.

phased array antenna: A steerable antenna that relies on electronics at the chip level to steer a signal rather than relying on motors to point an antenna.

Smart Register Platform (SRP): A software and networking application to allow survey data to be collected in the field and aggregated in a central location.

smart speaker: A type of voice command device and wireless speaker with an integrated virtual assistant that offers interactive capabilities.

smartphone: A mobile phone with similar functionality to a computer, typically having a touchscreen user interface, internet access, and the capability to run a wide variety of downloaded programs.

social score: An index reflecting a person's behavior or influence that can be used by individuals or organizations in evaluating trustworthiness, credit, or other factors.

spectrum: The range of wavelengths or frequencies over which electromagnetic radiation extends.

spyware: Software that enables a user to obtain information about another's smartphone or computer activities by transmitting data covertly from their device.

Stratolite: A telecommunications platform in the stratosphere, typically supported by a balloon or drone.

stratosphere: The layer of the earth's atmosphere between approximately 10 and 50 kilometers above the surface.

subscriber identity module (SIM): An integrated circuit, typically known as a SIM card, that stores the international mobile subscriber identity (IMSI) number and its related key. These are used to identify and authenticate users on a mobile system.

terabit per second (Tbs): One billion bits per second.

transmission control protocol / internet protocol (TCP / IP): A set of communication protocols that allows the connection of devices to the internet.

ultrahigh frequency (UHF): Radio spectrum frequencies between 300 megahertz (MHz) and 3 gigahertz (GHz).

Universal Basic Income (UBI): A financial service providing all citizens of a country or other geographic area with an unconditional fixed sum of money. The goal of the UBI is to reduce or prevent poverty and increase equality among citizens.

user interface (UI): Features designed into a device with which a person may interact. The UI can include keyboards, a mouse, display screens, and the appearance of a desktop.

V band: Frequencies in the microwave range of the electromagnetic spectrum ranging from 40 to 75 gigahertz (GHz).

virtual assistant: A software agent that responds to voice commands, speaks, and performs tasks or services.

Wi-Fi: A wireless local area network (WLAN) technology that connects electronic devices to each other and to the internet.

ACKNOWLEDGMENTS

I am deeply grateful to the many individuals and organizations that helped me explore The Great Connecting. Here is a partial list:

- **GAIA:** Todd Schafer, Joyce Jere, Nelson Khozomba, and many colleagues
- **Seeds of Learning:** Annie Bacon, Gail Chadwin, Patrick Rickon, Julián Ramón Guevara, Silvia Martinez, and many colleagues
- **The Hamels Foundation:** Heidi Hamels, Kathleen Greene, Alaina Baker, and colleagues
- **The Carr Foundation:** Greg Carr, Ryan Kirkham, Vasco Galante, and many colleagues
- **The Clinton Foundation:** Elizabeth McCarthy, Oluwaseun Aladesanmi, Marta Prescott, Nikhil Wilmink, Kalumbu Henry Pupe, and colleagues
- **Praekelt.org:** Debbie Rogers
- **Bluetown:** Nick Pallesen
- **Center for Social Integrity:** Aung Kyaw Moe, Jessica Olney, and colleagues
- **Global Ground Media:** Anrike Visser
- **Phandeeyar:** Thant Sin and colleagues
- **BRAC:** KAM Morshed, Azad Rahman, and colleagues

- **mPower:** Mridul Chowdhury
- **Internews:** Jeanne Bourgault, Michael Pan, Syed Zain Al-Mahmood, Hasan James, and many colleagues
- **CARITAS:** Sajal Debnath and colleagues
- **Loon:** Scott Coriell
- **Facebook:** Sara Su and Andy O'Connell
- **Gates Foundation:** Matt Bohan
- **Amazon:** Emily Roberts
- **Bamboo Learning:** Ian Freed
- **Stanford Business School:** Darius Teter
- **Harvard Kennedy School:** Nicco Mele, Phil Vermeer, Gene Kimmelman, George Twumasi, Dipayan Ghosh, Davan Maharaj, Suraj Yengde, Tom Wheeler, David Eaves, and many colleagues
- **Harvard School of Public Health:** Arlan Fuller and colleagues
- **Caribou Research:** Jonathan Donner
- **Forum One:** Chris Wolz, Dave Witzel, and colleagues
- **Readers:** Mark Couchman and Chris Grover
- **Editors:** Henry Carrigan and Kelli Anderson
- **Radius Book Group:** Mark Fretz and Evan Phail
- **Scribe Inc.:** Naomi Gunkel and the excellent team of editors and designers
- **Get Red PR:** Ann-Marie Nieves
- **Family:** My daughters, Maddie and Malia, whose enthusiasm is a gift, and my wife, Anne, my life partner in exploring everything

I was also assisted by countless people across Asia, Africa, and Latin America—development professionals, government employees, guides, drivers, and strangers—whose names I unfortunately don't know but whose generosity I greatly appreciate.

INDEX

ABOUT THE AUTHOR

Jim Cashel is a researcher and visiting fellow at the Harvard Kennedy School of Government. He also is chairman of Forum One, a web strategy and development firm. After completing studies in human biology at Stanford University and public policy and medicine at Harvard University, Jim has worked in international development, philanthropy, and technology.

When not on the road, he lives with his family in Sonoma, California.